INTRODUCTION TO SOCIOBIOLOGY

INTRODUCTION TO SOCIOBIOLOGY

Dr. Ravi Ranjan Kumar

Delve PUBLISHING

www.delvepublishing.com

Introduction to Sociobiology

Dr. Ravi Ranjan Kumar

Delve Publishing
224 Shoreacres Road
Burlington, ON L7L 2H2
Canada
www.delvepublishing.com

Email: orders@arclereducation.com

ISBN: 978-1-77469-855-6 (Hardcover)

ABOUT THE AUTHOR

Dr. Ravi Ranjan Kumar is currently serving as Head (officiating since 2021), Department of Social Work, Rajiv Gandhi Central University, Arunachal Pradesh. He has nearly a decade of teaching experience at Rajiv Gandhi Central University. An alumnus and Gold Medalist (2011) of Delhi School of Social Work, University of Delhi, his research interests include Education, Health, Governance, Political Economy, Social Justice and. Empowerment. Young Social Scientist Award by Indian Social Science Association, Research and Academic Excellence Award by Institute of Scholars, Karnataka, Gian Pandit Award and Kunda Datar Gold Medal by University of Delhi are some of his distinctions. The noted academician has over 40 Academic Publications which comprises of 22 Research Articles in reputed journals, a dozen of Book Chapters, and 6 Books (ISBN). He also has over two dozen of conference paper presentations apart from chairing and contributing to academic domains in capacity building and curriculum development as a resource person for a number of times. Other than academics, he also holds a distinction of serving Prasar Bharati (All India Rank -3) at All India Radio-National Channel (2015), and Deputy Superintendent of Police (2019), Bihar Police Service to mention a few.

TABLE OF CONTENTS

LIST OF FIGURES

LIST OF ABBREVIATIONS

ENA	Evolutionary Neuro-Androgenic
HBE	Human Behavioral Ecology
DNA	Deoxyribonucleic Acid
ABM	Account-*Based* Marketing
GDP	Gross Domestic Product
GPI	Genuine Progress Indicators
SSG	Sociobiology Study Group
RNAs	Ribonucleic Acid

PREFACE

This book introduces the readers to the concept and significance of sociobiology. This book sheds light on an overview of sociobiology, social evolution, population biology, social mechanisms, development, and modifications of social behavior, behavioral ecology, biological species as a higher form of existence, and ethical implications of sociobiology.

The first chapter introduces the readers to sociobiology. This chapter will emphasize on the history, theory, paradigm, instincts, and research methods involved in sociobiology. This chapter also explains primate sociobiology, primate social behavior and its development, and how sociobiology deals with social behavior.

The second chapter takes the readers through an introduction to social evolution. This chapter will provide highlights on early and later social evolution, multiplier effect in social evolution, dualities of evolutionary biology, and reversibility of social evolution. The chapter also explains parental care in animals and the four pinnacles of social evolution.

Then, the third chapter explains population biology. This chapter sheds light on population biology as a whole explaining population, population models, population heterogeneity, and population limitation and regulation.

The fourth chapter takes the readers through social mechanisms. This chapter introduces the mechanisms and problems of the confounders, concept of mechanism in social science, social mechanism as a typology, its elements and social dynamics.

The fifth chapter introduces the readers to the development and modification of social behavior. This chapter explains the behavioral modifications, social psychology, influence of social situation in creating social norms and environmental impact of economic growth. This chapter also gives a brief idea on models of link between economic growth and the environment, economic impact of environmental policy, and sustaining economic growth in the long term.

The sixth chapter throws light on the role of behavioral ecology in sociobiology. This chapter contains concepts on pop sociobiology, human behavioral ecology (HBE), and central criticism of Sociology and behavioral ecology.

The seventh chapter takes the readers through the concept of biological species as a higher form of existence. The readers are then told about organization, living entities, biological entities as self-reproducing units, altruism, and the meaning of sexual reproduction.

The last chapter of this book sheds light on the ethical implications of sociobiology. This chapter mentions about the origin of ethics, sociobiological ethics, and survival and its ideology, moral evaluation, moral norms, culture, and genes. The chapter also emphasizes certain topics like enlightened self-interest, ethical implications of evolutionary theory and altruism, and inclusive fitness.

This book has been designed to suit the knowledge and pursuit of the researcher and scholars and to empower them with various aspects of Sociobiology so that they are updated with the information. I hope that the readers find the book explanatory and insightful and that this book is referred by the scholars across various fields.

CHAPTER 1

Overview of Sociobiology

CONTENTS

1.1. INTRODUCTION

Sociobiology, a subfield of evolutionary biology, is the scientific investigation of the biological underpinnings of social behavior in animals, including humans. The scientific study of human societies and social behavior is known as sociology.

Sociobiology, not to be confused with the non-scientific school of thought known as "Social Darwinism," uses evolutionary theory to define and analyze patterns of social behavior as adaptations that emerge through natural and sexual selection.

Although sociobiological thinking is rooted in and related to older scientific disciplines such as ethology and comparative psychology, it emerged in 1975 as a new and distinct branch of behavioral biology with the publication of Sociobiology: Edward O. Wilson is an evolutionary scientist who specializes in myrmecology, or the study of ants.

Numerous detractors, including sociologists, protested towards the release of Sociobiology, claiming that it was simply one of the most current iteration of biological reductionism and biological determinism. However, by the mid-1980s, the sociobiological theory had gained traction in a number of behavioral and social disciplines, including psychology and anthropology.

At the same time, an increasing number of sociologists began to apply sociobiological reasoning to the study of phenomena such as marriage and family, ethnic relations, social inequality, crime, urbanization, and demography, and by the late 1990s, the term "evolutionary sociology" was being used to label the work of sociologists whose research is informed by evolutionary biology, particularly sociobiology and behavioral ecology.

Nowadays, some sociologists identify as "evolutionary sociologists," whereas others identify as "behavioral ecologists" or "sociobiologists." Despite certain ideological distinctions, all such evolutionary-minded sociologists share a dependence on the explanatory logic of the modern, neo-Darwinian scientific theory (Figure 1.1).

This article examines and describes the new style of evolutionary assessment that has arisen in sociology as a result of the influence of sociobiology as well as other evolutionary life sciences. The scientific study of the biological foundation of social behavior in animals, including humans, is known as sociobiology.

The new subject of evolutionary sociology arose from sociologists' recent efforts to include sociobiological reasoning in their theoretical thought and empirical study. Evolutionary sociologists oppose the Standard Social Science Model of human nature and social behavior in favor of the current model based on evolutionary biology.

Figure 1.1. E. O. Wilson–the creator of sociobiology gives advice to all young scientists.

Source: Image by Flickr.

The human mind, as well as intellect, are the result of natural selection development that happened in ancient, ancestral "environments of evolutionary adapted Ness." Early hominids overcome the restrictions to sociality provided by most ape species' weak social bonds and gained behavioral traits conducive to the establishment of main groups based on solid human connections.

Humans have acquired the ability to construct large-scale communities that are not limited to social bonds among genetic kin. Evolutionary sociologists discover and investigate social organization patterns that served as selection pressures in shaping the design qualities of the human brain and mind.

The human mind adapted to the size and organizational aspects of hunting-gathering civilizations, is confronted with unfamiliar social contexts in post-hunting-gathering times, particularly in urban industrial cultures.

Despite having substantial learning capacities and cultural resources, a "mismatch" between developed characteristics of human nature with different features of current, urban-industrial civilizations arises on occasion. This mismatch creates new behavioral issues for residents of urban-industrial civilizations, which manifest themselves in psychological, social, and health contexts.

Sociobiological theory and research findings have been used by evolutionary sociologists to design and pursue new study projects on conventional sociological issues such as sex, reproduction, family, kinship, social inequality, ethnic relations, and crime. During the second half of the twentieth century, 'sociobiology' developed as a new integrative field.

It changed the study of animal social behavior into a science-based on Neo-Darwinian logic, which views evolution as a change in gene frequencies in populations. This enabled the development of mathematical models that could be empirically tested, transforming conventional natural history into a more rigorous science.

However, the term "sociobiology" did never become a common term among practicing biologists. Many scientists engaging in genuine sociobiological research consciously avoided the name, preferring to refer to themselves as behavioral ecologists or functional ethologists.

This was partly due to the political overtones of sociobiology that arose as a result of the controversy surrounding Harvard scientist Edward O. Wilson's (1975) book Sociobiology: The New Synthesis. Those interested in what may be called "human sociobiology" were especially sensitive, preferring to be referred to as biosocial anthropologists, human ethologists, human behavioral ecologists, or Darwinian anthropologists. Several sociobiological research was renamed "evolutionary psychology" beginning in the 1990s.'

As this stood true, Wilson's specific, broadly defined sociobiological study program described in his book did not become the technique followed by practicing' sociobiologists' (although they did not call themselves that). Instead, the real sociobiological research paradigm in use was a modest but related set of theoretical ideas given by Richard Dawkins' The Selfish Gene.

'Sociobiology is the study of social processes inside the evolutionary biology conceptual framework. Its distinguishing feature is the interpretation of social events as outcomes of Darwinian processes of natural and sexual selection.

Sociobiologists use population biology, genetics, and ecology theories to predict and explain the evolution and diversity of social behavior within and between species, using the same 'adaptationist," selection logic that biologists use to understand evolution and diversity of anatomy and physiology. 'Adaptationist' refers to sociobiologists' general assumption that complex social phenomena such as mating preferences, offspring sex ratios, discriminatory treatment of young by parents, and so on have a functional significance for the behaving animal that can be illuminated by appropriate testing hypotheses. The term 'selectionist' refers to the belief that social traits have been fashioned by a history of selection to be successfully fitness promoting in ancestral contexts.

Although Darwin's contributions in the nineteenth century laid the groundwork, sociobiology did not mature until the 1960s, thanks to Hamilton's inclusive fitness theory and Williams' defense of the proposition that the differential reproductive success of individuals, rather than groups, is the level of selection that is most effective in creating adaptations.

The technique, often known as 'behavioral ecology,' has begun to dominate animal behavioral research. Sociobiology has added new perspectives to our knowledge of animal social behavior. It explains seemingly altruistic behavior in some animal species as genetically selfish because such behavior typically favors closely related animals whose genes mirror those of the altruistic person.

This understanding helps to explain why military ants lose their life to defend their colony, or why worker honeybees in a hive forego reproduction to assist their queen in reproducing. Sociobiology can explain variations in male and female behavior in some animal species as a result of the various methods the sexes must employ in order to pass on their genes to future generations. However, sociobiology becomes more contentious when it seeks to explain diverse human social behavior in terms of their adaptive significance for reproduction. According to one criticism, many of these behavior is more likely to be considered as cultural creations or evolutionary byproducts, having no clear adaptive function of their own (Figure 1.2).

Figure 1.2. Sociable weaver.

Source: Image by wikiwand.com.

Some sociobiologists, particularly Wilson, have been accused of assigning adaptive value to common but morally repugnant behavior (such as sexism and racism), so justifying them as natural or inevitable.

Sociobiologists argue that at least some parts of human behavior must be biologically driven (since competitiveness with other species will select for this feature); as evolutionary explanations of human behavior are not inherently flawed, but should be examined in the same manner as other scientific ideas, and that sociobiology doesn't really entail rigid biological determinism.

1.2. HISTORY OF SOCIO-BIOLOGY

Daniel Dennett, a philosopher of biology, proposed that Thomas Hobbes was the very first sociobiologist, claiming that in his 1651 book Leviathan Hobbes explored the origins of morals in human society from an amoral sociobiological viewpoint.

The animal behavior geneticist At a 1948 symposium on genetics and social behavior, John Paul Scott invented the term sociobiology, calling for the integration of field and laboratory studies in animal behavior research.

In 1956, the Ecological Society of America established a "Section of Animal Behavior and Sociobiology," which later became the American Society of Zoology's Division of Animal Behavior in 1958. Through his PhD student Stuart A. Altmann, who had been in regular touch with the members of the 1948 meeting, E. O. Wilson became acquainted with this nascent sociobiology in 1956.

Altmann was employed after developing his own brand of sociobiology to examine the social behavior of rhesus macaques using statistician 1965, and he was employed as a "sociobiologist" at Yerkes Regional Primate Research Center Wilson's sociobiology differs from those of the John Paul Scott and Altmann in that he drew on mathematical models of social behavior focusing on the maximization of genetic fitness developed by W. D. Hamilton, Robert Trivers, John Maynard Smith, and George R. Price.

The three sociobiology's by Scott, Altmann, and Wilson have in common the placement of naturalist studies at the center of animal social behavior research and the formation of alliances with emerging research methodologies, at a time when "biology in the field" was threatened by "modern" scientific practices (laboratory studies, mathematical biology, molecular biology).

Previously a technical word, "sociobiology" became generally recognized in 1975, when Wilson wrote Sociobiology: The New Synthesis, which provoked a heated debate. Since then, the term "sociobiology" has been widely associated with Wilson's worldview.

The book pioneered and popularized the attempt to explain the evolutionary mechanics of social behavior like altruism, aggressiveness, and nurturance, notably in ants and other Hymenoptera, but also in other species. However, biologists, and philosophers have been interested in the impact of evolution on behavior since the discovery of evolution itself.

Mutual Aid: A Factor of Evolution, authored by Peter Kropotkin in the early 1890s, is a prominent example. The book's last chapter is devoted to sociobiological research. Despite the public debate about its relevance to humans, Edward H. Hagen says in The Handbook of Evolutionary Psychology that sociobiology is "one of the scientific successes of the twentieth century."

"Sociobiology is now a key research and education component in practically all biology departments, and it serves as the framework for the work of nearly all field biologists." Sociobiological study on nonhuman species has grown substantially and steadily in the world's major scientific publications, including Nature and Science. To minimize public debate, the more generic term behavioral ecology is frequently used in place of sociobiology.

1.3. SOCIOBIOLOGY AS A FIELD OF STUDY

Sociobiology is indeed a branch of the evolutionary study of behavior that grew in popularity with the rise of ethology in the 1950s. Intensive fieldwork and theorizing over the next few decades resulted in descriptions of various animal societies in natural settings, a detailed understanding of the proximate experiential and physiological causes of behavior, a greater awareness of complex developmental processes, and new methods for assessing the evolutionary functions of behavior.

The integration of etiology and comparative psychology resulted in significant advancements in the study of proximal causes and development. The study of behavioral adaptation has benefited from the convergence of ethology, population biology, and ecology. This resulted in the development of behavioral ecology and sociobiology. The primary purpose of ecological research on social behavior is to investigate if differences in social structure and behavior can be explained by unequal selection forces. Because it is unusual that this topic can be answered by experimental, numerous correlational methodologies have been devised.

Since Darwin, pretty much the entire species comparisons have begun with the notion that commonalities can develop from one of two sources: a common ancestry or common selective forces. The natural selection theory is supported by analogous adaptations in unrelated species subjected to similar selective forces. When closely related species subjected to differing selection pressures demonstrate behavioral variations that make adaptive sense, this provides additional support. In a seminal study of monkey cultures, Crook, and Gartlan used this comparative technique. Ecological characteristics like food dispersal and predator pressure accounted for more similarities and variations in social structure than phylogeny.

Other animal species give significant convergent evidence that animal social structure can represent adaptations to ecological factors. Whenever there is in between-species variation in a feature, the evolutionary function may be addressed by comparing fitness between individuals with various trait values. Fitness, or reproductive success, is operationally defined as some measure of reproductively viable offspring contribution to the following generation (e.g., the number of weaned offspring). An early example compares black-headed gulls that did and did not nest in a colony (Figure 1.3).

Figure 1.3. Sociobiology–elephant males throughout their lives, challenge one another to establish dominance.

Source: Image by Flickr.

Those who nested in the colony had considerably more offspring, suggesting that colony nesting has an evolutionary role. This finding was bolstered by replication over time and the discovery of a plausible proximal mechanism: The colony nesters hatched simultaneously, which may have surpassed the capabilities of the local predators to consume all of the hatchlings.

1.4. SOCIOBIOLOGICAL THEORY

The modern synthetic theory was forged by combining Darwinian ideas with Mendelian genetics (e.g., Mayr, 1963). Two key elements of the synthesis are as follows: Natural selection is the driving force behind evolution, and evolution can be measured as changes in the relative frequencies of particular alleles within a population's gene pool. Darwin hypothesized natural selection to explain why evolution occurs.

Natural selection refers to environmental conditions that decrease the relative frequencies of some traits in a population because they lead to death or reproductive failure, which thus allows the relative frequencies of more successful traits to increase. Natural selection does not create new traits that would be adaptive. Novelty arises from chance mutations, each of which is assumed to make a small change.

Thus, adaptations arise gradually from the slow, cumulative honing of traits over generations by successive iterations of randomly generated change, followed by environmentally selected differential reproduction. Over time, a collection of traits emerges that fits environmental contingencies ever more closely. Heredity is necessary for adaptations to accumulate in a population.

Darwin knew that offspring resembled their parents, but he did not have a satisfactory theory of heredity. Mendel provided such a theory by linking specific traits with particulate units of heredity that passed unchanged from one generation to the next.

Although traits are affected by many genes during their development and single genes can affect the development of many different traits, the synthetic theory typically defines both traits and genes in a particulate manner. Thus, an inherited trait is considered to have a single gene as its heritable basis.

With this as an underlying assumption, evolution was redefined as changes in the relative frequencies of particular genetic variants (alleles) in the gene pool-the sum of all the alternative forms of all of the genes available to an interbreeding population. Sociobiologists are interested in how behavior can be explained logically as a result of selective pressures in the history of a species. Thus, they are often interested in instinctive, or intuitive behavior, and in explaining the similarities, rather than the differences, between cultures. For example, mothers within many species of mammals–including humans–are very protective of their offspring.

Sociobiologists reason that this protective behavior likely evolved over time because it helped the offspring of the individuals who had the characteristic to survive. This parental protection would increase in frequency in the population. The social behavior is believed to have evolved in a fashion similar to other types of nonbehavioral adaptations, such as a coat of fur, or the sense of smell.

Individual genetic advantage fails to explain certain social behaviors as a result of gene-centered selection. E. O. Wilson argued that evolution may also act upon groups. The mechanisms responsible for group selection employ paradigms and population statistics borrowed from evolutionary game theory. Altruism is defined as "a concern for the welfare of others."

If altruism is genetically determined, then altruistic individuals must reproduce their own altruistic genetic traits for altruism to survive, but when altruists lavish their resources on non-altruists at the expense of their own kind, the altruists tend to die out and the others tend to increase. An extreme example is a soldier losing his life trying to help a fellow soldier.

This example raises the question of how altruistic genes can be passed on if this soldier dies without having any children. Within sociobiology, a social behavior is first explained as a sociobiological hypothesis by finding an evolutionarily stable strategy that matches the observed behavior. Stability of a strategy can be difficult to prove, but usually, it will predict gene frequencies.

The hypothesis can be supported by establishing a correlation between the gene frequencies predicted by the strategy, and those expressed in a population. Altruism between social insects and littermates has been explained in such a way.

Altruistic behavior that increases the reproductive fitness of others at the apparent expense of the altruist, in some animals has been correlated to the degree of genome shared between altruistic individuals. A quantitative description of infanticide by male harem-mating animals when the alpha male is displaced as well as rodent female infanticide and fetal resorption are active areas of study. In general, females with more bearing opportunities may value offspring less, and may also arrange bearing opportunities to maximize the food and protection from mates. An important concept in sociobiology is that temperament traits exist in an ecological balance. Just as an expansion of a sheep population might encourage the expansion of a wolf population, an expansion of altruistic traits within a gene pool may also encourage increasing numbers of individuals with dependent traits. Studies of human behavior genetics have generally found behavioral traits such as creativity, extroversion, aggressiveness, and IQ have high heritability.

The researchers who carry out those studies are careful to point out that heritability does not constrain the influence that environmental or cultural factors may have on those traits. Criminality is actively under study, but extremely controversial. Various theorists have argued that in some environments, criminal behavior might be adaptive.

The evolutionary neuro-androgenic (ENA) theory, by sociologist/criminologist Lee Ellis, posits that female sexual selection has led to increased competitive behavior among men, leading to criminality in some cases. In another theory, Mark Van Vugt argues that a history of intergroup conflict for resources between men has led to differences in violence and aggression between men and women (Figure 1.4).

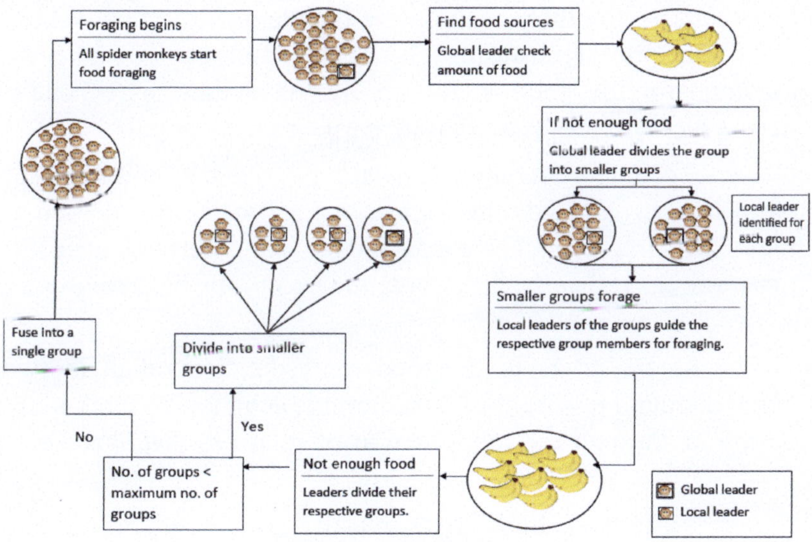

Figure 1.4. Social behavior of spider monkeys with respect to SMO algorithm is presented showing food foraging behavior of spider monkeys, reflections of fission-fusion structure and phases of algorithm.

Source: Image by Wikimedia commons.

The novelist Elias Canetti also has noted applications of sociobiological theory to cultural practices such as slavery and autocracy.

1.4.1. The Unit of Selection

Sociobiological theory has been most concerned with two related issues:

- specifying the unit of selection for social behavior; and
- dealing with the apparent anomaly of altruistic behavior.

Altruistic behavior is defined as behavior that reduces individual fitness while it contributes toward the fitness of another. Such behavior is relatively commonplace, even if one excludes parental care. For example, alarm calls, which expose the caller while alerting con-specifics to danger, are widespread. And some animals delay or forgo reproduction while they help to care for the offspring of others.

Wynne-Edwards (1962) proposed that such individual sacrifice evolved for the good of the group, and criticism of group selection (Williams, 1966) served as a focus for the emergence of sociobiological theory.

The basic criticism was that an individual whose behavior reduced its own fitness relative to others in the same population could not accumulate genes in the gene pool. True altruism is not an evolutionarily stable strategy. Behavior done for the good of the group but at the cost of one's own reproductive success would soon disappear from most populations.

Yet, some behavior certainly appears to reduce individual fitness. Hamilton (1964) resolved this apparent anomaly by demonstrating that fitness-reducing behavior could indeed arise through selection on individuals, provided that the "inclusive fitness" of the behaving individual was enhanced.

Inclusive fitness refers to the total representation of an individual's genes in the population gene pool, whether it is achieved through individual reproduction or through the reproduction of relatives that share the same genes.

Therefore, altruistic behavior directed toward offspring, siblings, parents, and even more distant kin can arise through the action of natural selection on individuals. Observed from the perspective of fitness, such altruistic behavior is often described as selfish.

Unfortunately, it is sometimes forgotten that in this context both altruism and selfishness are metaphors, not motives. Hamilton's resolution of the altruism problem requires nepotism. Thus, selection through inclusive fitness has been called "kinship selection." Many animals do live in kinship groups, and some discriminate kin from nonkin as recipients of help (Sherman & Holmes, 1985).

Trivers (1971) showed that through reciprocity, altruism can also evolve among unrelated individuals. Reciprocal altruism is not fitness-reducing when altruistic performers (or genetic relatives) can expect to receive equivalent aid at another time.

The assumptions of kinship selection and reciprocal altruism, like those of optimality and game theory modeling, add complicating layers to the phenotype. Of course, animal behavior can be very complex. Some animals engage in probability matching, others keep track of large matrices of individual relationships, and so on.

Nevertheless, according to the synthetic theory of evolution and sociobiology, natural selection ultimately modifies allele frequencies in a population by the differential reproduction of individuals carrying particular alleles. The theories thus assume one allele for each adaptive phenotype, however complex.

A successful theory of evolution must specify the relation between units that are inherited and units that are selected, but what are these units and how are they related? Sociobiological theory identifies genes as the unit of heredity.

It also rules out the group as a unit of selection, because it is impossible for one of two or more alternative phenotypes to increase in frequency if the individuals expressing the phenotype decrease their individual (inclusive) fitness, regardless of how beneficial the group as a whole might find the behavior. In an influential book, Williams proposed that genes are the most likely unit of selection.

There is some ambivalence within sociobiological theory over whether selection can sometimes act on whole genotypes, chromosomes, or something larger than a single gene, but the selected unit has been moved from individuals as well as from groups and lodged in the genetic material.

Dawkins (1976, 1982) argued that organisms and their phenotypes are just vehicles used by genes to make more genes. He reasoned that evolution requires units with the capacity for faithful self-replication. Sexually reproducing individuals do not replicate themselves. Individual genomes undergo meiosis and recombination.

Chromosomes, which can self-replicate, are frequently rearranged by crossing-over and other processes. Dawkins defined a gene as a length of deoxyribonucleic acid (DNA), perhaps only as long as a cistron but possibly much longer. To respond successfully to natural selection, such a gene must maintain its integrity for a long time, and it must copy itself faithfully and frequently.

According to Dawkins, the unit of heredity and the unit of selection are one and the same. Thus, with respect to altruism, Dawkins postulates a single gene that accounts for both the altruistic behavior and the recognition of an appropriate recipient.

However, genes do not behave; animals do. Because animals survive or die, reproduce or not, based on their environmental fit, genes must be picked indirectly. Dawkins agrees that behavior and other complex traits are the results of many genes and many nongenetic variables working together.

In this sense, he defines genes in terms of how they affect the natural selection, that is, the difference they make between people, rather than how they generate phenotypes. "Creating a leg is a multi-gene collaborative effort. External environmental influences are also necessary: after all, legs are composed of food! But there might be a single gene that, everything else being equal, causes legs to be longer than they would have been without the impact of the gene's allele.

Dawkins, on the other hand, views genes as phenotypic development directors. Organisms are defined as machines designed by genes to help them survive. "Every action made by a survival machine is a risk, and it is the duty of genes to program brains in advance so that they make decisions that pay off on average." Genes, according to Dawkins, not only distinguish people but also carry blueprints for phenotypes. This is not a minor change in the definition.

In sociobiology, genes are frequently seen as blueprints, information, or guidance for phenotypes. It agrees with the definition of evolution as changes in the relative frequency of alleles in a gene pool. However, as the developmental sciences have demonstrated, genes are not blueprints.

Genes, on the other hand, are physical components of cells that, together with other components, produce proteins. Their operation is heavily influenced by internal agents (including those that enter the cell from adjacent cells), distant sections of the body, and external factors. There seem to be numerous layers of separation between the activities of genes within cells and the overall function of the organism in its environment.

More levels are added through social interactions, connections, and group structures. The number of stages traversed by each species and the intricacy of their structure at each level varied.

One of the most fundamental and crucial biological concerns is how these levels are bridged. However, sociobiology is a level less theory. There are simply genes and phenotypes, with the assumption of a mapping connection between them, so that each distinct phenotypic corresponds to a different gene and each alternative phenotype corresponds to an alternate allele.

Organisms are reduced to mosaics of discrete adaptive features when stripped of their organization, and genomes are reduced to similar mosaics of individual genes; each gene is responsible for an adaptation. Sociobiologists investigate the adaptability of behavioral phenotypes, rather than their antecedent causes, by putting hypothetical phenotypes into models and comparing the findings to real phenotypes recorded in the lab and in the field.

Sociobiologists are more interested in individual differences than in how phenotypes evolve. Nonetheless, the theory must handle development since evolution is portrayed as gene selection, with genes competing by creating better phenotypes. In terms of development, genes are portrayed as the directors who generate certain behavioral features. The notion that genes serve as predetermining, predisposing, or directing factors contradicts results in developmental genetics, developmental biology, and developmental psychobiology (Figure 1.5).

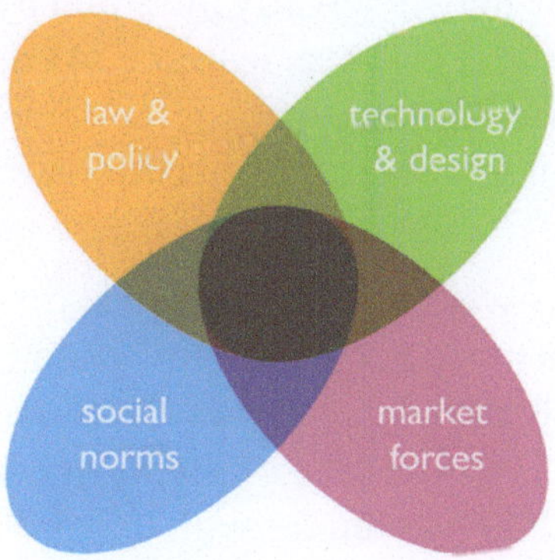

Figure 1.5. Four mechanisms that regulate our behavior.

Source: Image by Flickr.

It can also be misconstrued to indicate that genes set imperatives for conduct that is extremely difficult to escape or that may endanger an individual if avoided. This perspective shifts biology from science to public policy, or as Kitcher phrased it, from scientific sociobiology to popular sociobiology.

1.5. SOCIOBIOLOGY DEALS WITH SOCIAL BEHAVIOR

To various people, sociobiology means different things. Sociobiology, as defined by scientists, is the study of social behavior and its development. Sociobiologists typically explain behavior in terms of evolution. Although sociobiologists are more concerned with the genetic components of behavior, they all accept that learning may modify much of behavior.

They also accept that natural selection can only affect inherited components of behavior. As a result, one of the primary studies aims of sociobiology has been the examination of the relative relevance of acquired and hereditary impacts on certain behaviors.

A number of critics of sociobiology question the emphasis on inherited behavioral traits. These detractors would rather stress learning, especially cultural learning in people, as a powerful effect on conduct. We shall investigate both points of view.

1.5.1. Learned and Inherited Behavior

Many behavioral patterns may be greatly modified by experience in dealing with the environment, often known as learning. Almost every behavior that has been thoroughly studied has a genetic component. Although learned behavior can improve fitness, natural selection can only alter the genetic components (or predispositions) underlying the behavior.

Natural selection may work on the ability to learn specific types of things, such as how to navigate a maze of one's environment. In such circumstances, the quality valued by selection is not the action itself, but rather the ability to learn the behavior. This is true of one of the most commonly studied forms of learnt behavior: the ability to run through mazes.

The capacity of rats to learn specific mazes was evaluated, and the number of training sessions required for the rats to learn the mazes was noted. Their littermates were subsequently selectively bred for numerous generations despite never having been examined.

Breeding the littermates of fast maze-learners produced a strain with a low average number of training sessions required, whereas a strain of slow learners was produced by breeding the littermates of animals who required more repetitions. The use of littermates in this experiment minimized the role of learning experience or other effects in determining the differences between the two strains.

It is worth noting that the behavior was not entirely governed by genetics; it had to be learnt. This behavioral feature is determined by a combination of genes and environmental effects. The difference between the two strains was caused by the accumulation of gene combinations, which was only feasible since some variance between groups was genetically determined (Figure 1.6).

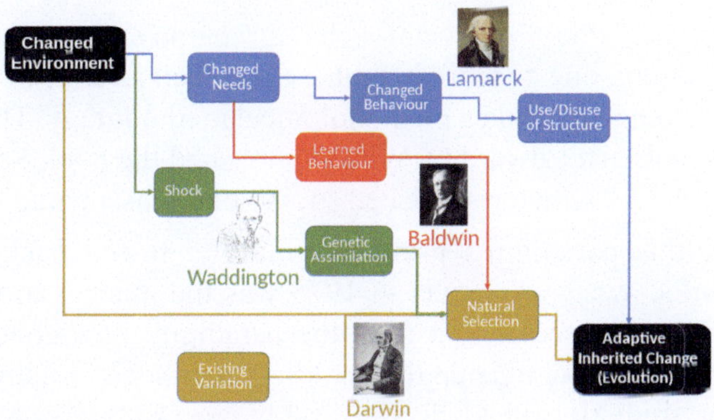

Figure 1.6. Lamarck's theory of evolution compared to Darwinian evolution, Baldwin effect, and Waddington's genetic assimilation.

Source: Image by Wikimedia commons.

Additionally, saying that group variation is heritable does not imply that the behavior is inherited as a permanent and unchanging feature. There are very few behaviors in any animal (and none in humans) that are not modifiable by learning. Nobody can learn to play basketball like Michael Jordan or the cello like Yo-Yo Ma without years of practice.

Nobody can become even a poor basketball or cello player without a significant amount of practice—a learning phase. However, some intrinsic aptitude and ability are undoubtedly required; otherwise, any of us might become a brilliant basketball player or a world-renowned cellist just by practicing hard. As a result, it is critical to underline that the frequently asked subject of learnt vs. hereditary behavior is a false dichotomy.

Every taught behavior is based in part on some innate learning capacity, which may include the ability to learn some types of behaviors but not others, to respond to certain stimuli but not others, to learn to a particular level of complexity, and so on. Similarly, most inherited behavioral patterns may be adjusted to some extent by learning.

These findings support the testable premise that almost every behavior pattern is at least partially learnt and partially inherited. Innate behaviors are those that do not require learning, and they are thought to have a hereditary component in any species, no behavior is entirely taught, and only a handful are entirely inherited.

1.6. THE PARADIGM OF SOCIO-BIOLOGY

Sociobiology, the study of social behavior among species, employs a scientific paradigm, one or more hypotheses, as well as a set of value-laden assumptions, a language, and a methodological methodology. The American evolutionary biologist Edward O. Wilson published the book Sociobiology: The New Synthesis, which established sociobiology as a paradigm.

Many of this paradigm's principles may be traced back to Charles Darwin's works. What was novel in 1975 was the manner in which these concepts were combined to construct the paradigm. Few concerns would have been made if those outside the paradigm saw sociobiology as nothing more than the study of social behavior.

However, sociobiology has been widely chastised for its emphasis on hereditary behavior. Many critics of sociobiology have noted that most of the behavior is learnt, and nearly all behavior, particularly in mammals, may be changed by learning.

Furthermore, because language and culture have a substantial impact on human behavior, many scientists who are generally supportive of sociobiology have cautioned against extending sociobiological discoveries from those other animals to people. Certain non-scientific critics of sociobiology are concerned about genetic determinism, the belief that our particular qualities are predetermined at birth and cannot be modified.

As we explored in chapter, genetic determinism is feared because individuals in power have attempted to dominate others (other socio-economic classes, races, and women) throughout history by preaching that existing inequities were 'natural,' based on fundamental and immutable differences. Furthermore, many individuals are concerned that simply asserting that behavior is intrinsic may prevent people from attempting to modify that behavior via education or other ways. The argument that conducts is intrinsic can be especially dangerous to social reformers who base their expectations for the future on people's ability to change their behavior. Those who believe in genetic determinism are in the minority among scientists. Most biologists, particularly those who research animal behavior, are astounded by the extent to which behavior may vary in response to environmental factors, including the conduct of other people.

There are genetic boundaries on what can and cannot be taught, yet the behavior is astonishingly adaptable (or 'plastic') in most animal species within these limits. No behavior is completely 'controlled' by either heredity or environment—almost every behavior is impacted by both of these variables throughout the individual's lifespan.

1.7. RESEARCH METHODS IN SOCIO-BIOLOGY

No approach can examine behavior unless it has been fully explained. As a result, sociobiology comprises a large number of animal observational field investigations. How can one tell the difference between learnt and intrinsic components of a behavior? Sociobiologists explore these components using the methods listed below.

1.7.1. Rearing Animals in Isolation

Animal isolation is a common form of experiment in which an animal is raised in isolation, in a soundproof room with bare walls and little possibility for learning, including no possibility to learn behavior from others. The animal's behavior under these settings is thought to be mostly intrinsic. Such experiments on humans are not morally permissible.

1.7.2. Rearing Animals under Different Conditions

If the same activity is performed in the same way by animals or humans raised under drastically different conditions, the behavior is mainly intrinsic. If, on the other hand, the behavior changes according to the circumstances of raising, the variance can be attributed to environmental impacts, however, this does not rule out hereditary influences. Cross-cultural studies examine the actions of persons reared in different communities or conventions; intrinsic tendencies are believed to be consistent across cultures, whereas taught behavior patterns are predicted to vary.

1.7.3. Studying Behavior in Different Genetic Strains

When different strains or breeds of a species differ behaviorally in a consistent and recognized way, a significant genetic component emerges. (This does not rule out the possibility of learnt components being there).

1.7.4. Conducting Adoption Studies

If two groups vary in a certain habit, it may be interesting to research individuals from one group who are adopted and reared by the other. Under these settings, conduct that consistently resembles the birth population reveals a hereditary influence, whereas behavior that resembles the rising population demonstrates a taught impact. Mixed or contradictory findings may suggest the presence of both factors.

1.7.5. Conducting Twin Studies

If a characteristic is strongly influenced by genetics, identical twins should display the feature anytime either one does, but fraternal twins are more likely to diverge.

Human twin studies are widely challenged because the effects of learning cannot be clearly distinguished from those of inheritance unless the twins are nurtured separately in randomly selected households, settings that are seldom even approximated. Some studies compare identical twins raised together versus those raised apart, an experimental design that seeks to overcome some of these challenges.

1.8. INSTINCTS IN SOCIO-BIOLOGY

Instincts are a subset of inherent behavior. Instincts vary from other inherent actions in that they are sophisticated behavior patterns with strong genetic control. The standard test for determining if a behavior is instinctual is whether it arises at the proper stage of life in an animal raised in isolation since birth or hatching.

For example, if a songbird raised in a soundproof chamber performs the species and sex song when it reaches adulthood, the song is deemed instinctive. Numerous behaviors investigated in fish, birds, and many invertebrates (including insects) have been found to be essentially instinctual using this technique. Most species have substantial innate components to courting and mating behaviors. Other instinctual behavior patterns include automatic 'escape' conduct, nest-building activity, orb-weaving in spiders, and other threat gestures. When instinctual behavior leaves a lasting product like a nest or a spider's web, these products are typically so unique that the species that generated them may be identified.

Mammals, generally, rely on learnt behavior rather than instinct. Many actions that are innate in other animals have substantial acquired components in primates. These habits may differ substantially amongst human groups.

1.8.1. Advantages of Instincts

Animals with short lifespans rely mainly on instincts. Mayflies (insects of the order Ephemeroptera), for example, have an adult life span of fewer than 24 hours. They do not eat during this brief period, but have only enough time to find a mate, copulate, lay their eggs, and die.

There is no time for learning, and there is no time for mistakes. Mayflies that complete their mission effectively are those that can conduct their behavior accurately on the first try; they are unlikely to receive a second opportunity.

As a result of millions of years of selection, adult behaviors have become instinctual and routine, leaving no possibility for variation or creativity this is true of instincts in general: behavior is instinctive when uniformity and automatic reaction are adaptive, whereas creativity and variation may be detrimental.

If the behavior is intuitive, a simpler brain allows for more complexity; acquired behavior of equivalent complexity necessitates a more intricate neural system as well as a lengthy learning phase during which numerous mistakes have been made.

1.8.2. Mating Behavior

Courtship (attracting a mate and getting accepted as a mate) and the actual release or transfer of gametes are both examples of mating behavior. Except for higher primates, virtually all animals have a significant innate component to their mating behavior.

Scientists can show the instinctual component of most types of mating behavior by growing people in isolation till sexual maturity and then testing them to determine if they can conduct the behavior typical characteristic of their species. Natural selection promotes consistency over variability in mating behavior.

In several animals, such unvarying behavior (known as stereotyped behavior) is employed for mate finding and recognition. Each species' behavior is developed to fit the sort of signal that each can detect, such that visual mating signals are utilized by species with good eyesight, chemical signals by species with good chemical reception, and sounds by species with high sound discrimination.

Many birds, frogs, and insect species employ sounds as mating signals, and the non-calling sex (typically female) reacts exclusively to mating calls of the correct pitch, length, and pattern of repetition. Mating signals and the behavioral responses to them are both instinctual (Figure 1.7).

Figure 1.7. Elephant mating ritual.

Source: Image by Wikimedia commons.

Participants of each sex are aware of what to look for in the other sex and normally shun nonconformists who depart from the natural pattern. Nonconformists are consequently penalized by sexual selection since they rarely mate and so have no progeny. Firefly flashing patterns, while visually appealing, are sexually picked in the same way.

Mating sounds and visual displays are finely regulated within a small range for each species due to sexual selection. Closely related species' mating sounds and wooing practices frequently differ. Differences in mating calls and other wooing displays are frequently used as reproductive isolation strategies to keep species from interbreeding.

During courting, male birds of several species show brightly colored parts. Beautiful, ornate mating rituals arise as a result of sexual selection in animals where the discerning sex (the one performing the choosing) regularly favors the most visible displays. Peacocks, lyrebirds, and birds of paradise are well-known for their elaborate male plumage.

Male birds having less prominent plumage may instead focus on creating an ornate nest. South Pacific bowerbirds construct their nests within a huge framework (a bower) that also functions as a dating site. A few species even construct an 'avenue' paved with colored stones leading to the bower's entrance (Figure 1.8).

Figure 1.8. Sarus cranes mating behavior.

Source: Image by Flickr.

Bowerbirds with ornate plumage, in general, do not construct complex bowers, and the species that build impressive bowers using simple plumage.

1.8.3. Territorial Behavior

Across many animals, one or both sexes may exhibit territorial behavior by guarding a region all year or only during mating season. Many animal species defend their territories against invaders of the same kind. Only males in certain animals are territorial. Territorial behavior separates individuals and pushes losers to seek out new areas, hence increasing the species' range whenever possible.

Each territory must include enough food for a mating couple and their progeny, as well as places to hide and seek safety and at least one acceptable nest site. Males who lack a territory are frequently unable to attract partners and so produce no progeny during that season.

The precise limits of territory and one's position as a territory holder or trespasser are taught, but the overall inclination to form territories in some species (or sexes or seasons) is natural. Territorial species might threaten territorial rivals through gestures (Figure 1.9).

Figure 1.9. Trumpeter swan territorial aggression.

Source: Image by Flickr.

Mammals that form territories may use their own smell to designate their region. The intimidation of rivals by gestures or the presence of scents serves to separate persons without inflicting violence or death. Such ritualized forms of territorial protection are far more widespread than any sort of combat that is likely to result in harm.

1.8.4. Nesting Behavior

The selection of a nesting location may be a key aspect of territorial behavior. As part of the mating ritual in certain bird species, the male constructs the nest and then presents it to the female. In some species, the male and female may work together to build the nest as part of the mating ritual.

Figure 1.10. Photo of Canada goose nesting on beaver lodge in Crawford county, Pennsylvania.

Source: Image by Wikimedia commons.

Females can incubate the eggs on their own, although males can help by providing food or guarding the location from predators. Males and females in other species alternate protecting the nest and sitting on the eggs. Feeding the hatchlings can either be a lonely or collaborative effort (Figure 1.10).

Individuals engage in the actions just mentioned groups of creatures can also execute behaviors, which we will discuss in the next section.

1.9. PRIMATE SOCIO-BIOLOGY PRESENTS ADDED COMPLEXITIES

Monkeys, apes, lemurs, tarsiers, and humans are members of the primate order. All primates are incredibly sociable creatures. They are so eager to engage with other members of their species that they will go to considerable lengths to sustain such interaction or simply to gaze. We can design an experiment to put this idea to the test.

Set up a partition that totally blocks the view through a window, and gives a lever that lifts the barrier for a specified amount of time, allowing you to see through the window temporarily. Most primates will then spend hours pressing the lever and peering through the window, then returning to press the lever for another glimpse practically as soon as the barrier lowers.

Whenever the window shows moving items (such as electric trains) rather than non-moving ones, the pace of lever pushing increases (such as furniture). The rate is even greater if an active moving animal is visible through the window, and it is even higher if the view includes additional primates of the same species.

Is it any surprise that individuals spend hours peering out windows at their surroundings, especially when other people's emotions and interactions are visible? In addition to the live activity visible via a physical window, television offers a virtual window through which we may see people interact even more closely.

1.10. PRIMATE SOCIAL BEHAVIOR AND ITS DEVELOPMENT

Social abilities in both humans and nonhuman primates are heavily reliant on early life learning. All parents, current, and prospective, should be aware of the critical relevance of early childhood experiences in all facets of human existence. One of the crucial areas in which humans and other primates are quite similar is the enduring relevance of learning that occurs very early in life.

Because all primates are so similar, data from nonhuman primate trials are frequently utilized to obtain insights into human behavior, albeit findings from one species should not be generalized naively to another. When comparing the behaviors of different monkey species, however, we generally discover far more resemblances than differences.

1.10.1. Early Development of Behavior

As previously noted, the usual test for an instinct needs an animal to be reared in isolation. However, raising a monkey in solitude resulted in deviant behavior similar to that of mistreated children. According to Sigmund Freud, a baby's connection to its mother is first motivated by the desire for feeding.

Harry Harlow of the Wisconsin Primate Research Center tested this concept by raising young rhesus monkeys (Macaca mulatta) with various sorts of care but with no live moms. Instead, dummy 'mothers' with brightly colored wooden 'heads' carried infant bottles in wireframes.

Although the baby monkeys drank the milk, their behavior became increasingly aberrant over time. The infants were easily scared and regularly cowered in the corner. They developed no emotional relationships and appeared to disregard their mothers except when they were hungry.

Because the young monkeys failed to treat the wire model as a mother, Freud's hypothesis was proven false. For newborns to build a link with their mothers, something more than milk was required.

1.10.2. Development of Adult Behavior

Rhesus monkeys reared with terrycloth mothers tend to operate properly until sexual maturity when behavioral abnormalities occur. A typically reared male rhesus monkey mounts' a female throughout her reproductive cycle if she 'presents' to him, however, the motherless males never mounted any females and appeared to be clueless in this circumstance.

Motherless females went through their reproductive cycles (their genitals enlarged and became brilliant pink), but they never 'presented' to any test males and persistently refused any sexual approaches. A handful of these women were artificially inseminated and became pregnant. When their children were born, they displayed no maternal behavior, such as Picking up their infants and holding them to their breasts is one example. Instead, parents neglected or rejected their children, in some cases so forcefully that the newborns were removed for their own safety.

Sexual and parental behavior had never been learnt in these monkeys, despite the fact that their conduct had appeared normal up to the time of sexual maturity. Adult social behavior in rhesus monkeys and other higher ape species is heavily influenced by learnt components.

1.10.3. Rough and Tumble Play

Most primate play is "rough and tumble," with regular and repetitive physical contact, including pushing, tugging, and climbing—observe young children in a playground, for example. Primate play also includes chasing and avoiding, which is frequently followed by rough and tumble play.

Although rough and tumble play is neither sexual nor maternal, it appears to offer a variety of lessons, including how to manage and maybe restrict other people without injuring them. Hurting another person, whether intentionally or unintentionally, results in squeals of agony, prompting parents to interfere and break up the action.

Play also teaches taking turns in various roles such as pursuer and chased, restrainer, and restrained, climber, and supporter, and so on. In the context of a game, The participants learn how strong or weak other people are, as well as how much physical play they can handle.

1.11. CONCLUSION

Sociobiology, a branch of evolutionary biology, is the scientific study of the biological bases of social behavior among animals, including humans. In the years immediately following the publication of Sociobiology, numerous critics, including sociologists, objected that sociobiology was but the most recent iteration of biological reductionism and genetic determinism.

Sociobiology is more controversial, however, when it attempts to explain various human social behaviors in terms of their adaptive value for reproduction. Sociobiology is the scientific study of the biological bases of social behavior among animals, including humans.

Those involved in what might have been called 'human sociobiology' were particularly sensitive, preferring to be called biosocial anthropologists, human ethologists, human behavioral ecologists, or Darwinian anthropologists. 'Sociobiology' is the study of social phenomena within the conceptual framework of evolutionary biology. This was large because of the political connotations of sociobiology caused by the debate around Harvard zoologist Edward O. Instead, the actual sociobiological research paradigm-in-use consisted of a quite small but connected collection of theoretical insights introduced by Richard Dawkins' The Selfish Gene.

Not to be confused with the non-scientific body of thought known as "Social Darwinism," sociobiology uses evolutionary theory to describe and analyze patterns of social behavior as adaptations that evolve by natural selection and sexual selection.

'Sociobiology' emerged as a new integrative discipline during the second part of the twentieth century. Sociobiology has contributed several insights to the understanding of animal social behavior. The name 'sociobiology,' however, did not become the name in use for practicing biologists.

REFERENCES

1. (n.d). *Sociobiology*. [e-Book] http://abacus.bates.edu/. Available at: http://abacus.bates.edu/acad/depts/biobook/ch08.pdf (accessed on 27 June 2022).
2. Ahad, M., (2014). *Sociobiology is Not a Theory of Evolution but a Branch of Entomology, which Deals with Social Insects*. [e-Book] https://www.researchgate.net. Available at: https://www.researchgate.net/publication/266794006_Sociobiology_is_Not_a_Theory_of_Evolution_but_a_Branch_of_Entomology_which_Deals_with_Social_Insects (accessed on 27 June 2022).
3. Daly, M., & Wilson, M., (2001). Sociobiology: Overview. *International Encyclopedia of the Social & Behavioral Sciences*, 14539–14543.
4. Encyclopedia Britannica, (n.d). *Teratology*. [online] Available at: https://www.britannica.com/science/teratology (accessed on 27 June 2022).
5. Lumsden, C., (2011). Sociobiology. *Encyclopedia of Creativity*, 352–359.
6. Machalek, R., (n.d). *Historical Developments and Theoretical Approaches in Sociology* (1st edn.) [e-Book] Encyclopedia of life support systems (EOLSS). Available at: https://www.eolss.net/sample-chapters/C04/E6-99A-14.pdf (accessed on 27 June 2022).
7. Moore, C., & Michel, G., (1998). *Sociobiology*. [e-Book] Garland. Available at: https://www.researchgate.net/publication/308515317_Sociobiology (accessed on 27 June 2022).
8. Wilson, E., (1975). *Sociobiology the Abridged Edition*. [e-Book] The Belknap Press of Harvard University Press. Available at: https://teoriaevolutiva.files.wordpress.com/2014/02/edward_o-_wilson_sociobiology_the_abridged_editbookos-org.pdf (accessed on 27 June 2022).

CHAPTER 2

Social Evolution: Introduction

CONTENTS

2.1. INTRODUCTION

In evolutionary theories, social evolution is analyzed and explained as a process of directional social change. Social evolution theories date back to Spencer, Morgan, Tylor, and Marx & Engels in the second half of the nineteenth century.

After a lapse, evolutionary theorizing revived in the 1930s and 1940s with the work of Childe, White, and Steward, and continued into the 1960s and 1970s with the work of Sahlins, Service, Carneiro, Lenski, and Harris. Important typologies of stages of evolutionary development have been developed by most of these thinkers.

The Neolithic Revolution and the rise of civilization and the state are two extremely important transformations in social evolution, despite the fact that there is no complete consensus on the most important dimensions. The former began about 10,000 years ago, the latter around 5,000 years ago, and both occurred on an independent basis in several regions of the world.

The rise of the capitalist world economy in the sixteenth century is regarded by many as the third and most recent great social transformation. Others regard the Industrial Revolution of the eighteenth century as the third great transformation.

Recent decades have seen rapid development in the area of social evolution theory, especially in the study of reproductive skew and selection in viscous populations. However, theoretical advances in behavioral ecology have far outpaced the empirical testing of models as is often the case. The result is a range of models each based on different assumptions which make distinct or overlapping predictions.

There have been over 15 different models published on the study of reproductive skew in the period 1998 to 2003 (Gardner & Foster, 2008); over 20 models have been published on the study of cooperativity in viscous populations since Taylor's (1992) first analysis using an infinite island framework (Lehmann & Rousset, 2010).

In the future, empirical studies should be conducted to uncover mechanisms or assumptions that do not hold up in nature and propose new mechanisms or assumptions that can form the foundation for theoretical work based on experiments and observations.

As for reproductive skew theory, our results demonstrate how threats of departure, eviction, or attack (or the lack thereof) determine reproductive distribution, and the likelihood of aggressive behavior. Group membership benefits are so high that threat of dispersal is irrelevant; threats of eviction and infanticide are usually ineffective as they are limited by information, which means that transgressors often escape punishment (Figure 2.1).

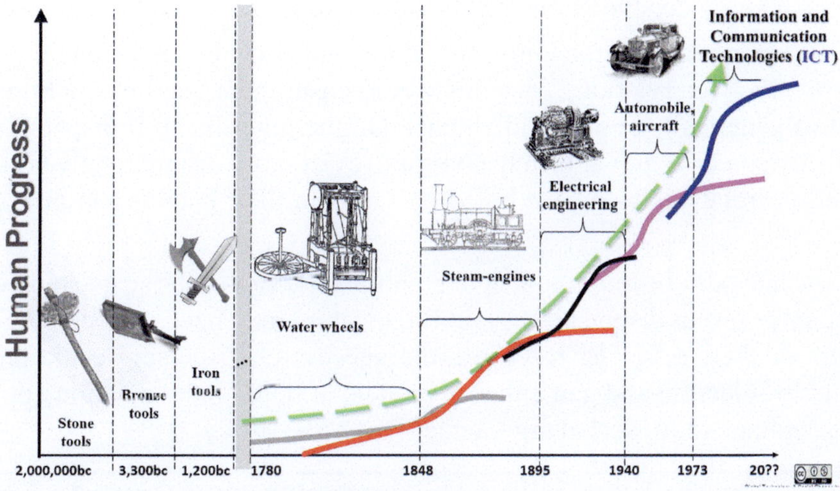

Figure 2.1. Long waves of social evolution.

Source: Image by commons.wikimedia.org.

These social groups are dynamic because dominants are unable to deter subordinate reproduction using threats, which results in large fluctuations in group size, mass evictions, group extinction, and group founding. To understand population dynamics of banded mongooses, and cooperative mammals more generally, requires knowledge of the behavioral strategies employed in reproductive conflict, and the physical and informational constraints on those strategies.

We believe that in many other cooperative vertebrates, there will be strong links between reproductive conflict, gene-flow, and population dynamics. Hopefully, a second wave of research on skew and reproductive conflict is forthcoming, grounded, and informed by long-term studies of cooperative breeders.

For demographic models of social evolution, researches on the banded mongoose and other cooperatively breeding vertebrates provide opportunities to test existing models, but also suggest areas where new theory is required. In particular, more work is needed to understand the causes and consequences of forced eviction from viscous groups—when should subordinates leave versus wait to be pushed?

In many cases dispersers could gain from leaving together as a group, but require third party intervention to solve the coordination problem of who leaves first. From the perspective of evictors, which individuals are targeted for dispersal may depend on the mortality rate or success rate of dispersers.

Where dispersal is sufficiently easy, dominants may be selected to target closer relatives for eviction, since this would ensure that budding coalitions consist of genetically similar individuals. On the other hand, if dispersal is sufficiently costly, dominants may do best to evict more distant relatives and leave closer relatives to inherit breeding status in their natal group later in life.

There may also be a component of positive feedback in the size of evicted groups: once group dispersal is common, evictors may need to expel larger numbers of their offspring to ensure the success of dispersing coalitions. Formal kin selection and game theory models of budding by eviction could help to evaluate these verbal arguments.

> Researchers studying evolution in structured populations suggest that behavioral ecologists studying cooperative breeders might need to reconsider their concept of fitness and how it is measured in natural populations. In structured populations, fitness is a measure of the relative success of with which group members can export copies of their alleles to other groups, by producing dispersers which usurp existing groups or found new ones.

This type of fitness concept is sometimes referred to as "metapopulation fitness" because each group (or deme) in the population can be thought of as a miniature population embedded within the metapopulation. For example, Metz and Gyllenberg (2001) suggest defining fitness in structured populations as "the average number of newborn dispersing individuals resulting from the local colony founded by a newborn disperser."

In a system such as banded mongooses, fitness might be similarly defined as the average number of emigrants produced by a successful group founder or usurper. These definitions are very different from the usual measures of fitness employed in field studies, namely the average or lifetime number of surviving offspring (Figure 2.2).

In long-lived cooperative breeders, selection on fertility and mortality, and on social behavior across the lifespan, depends on the age and sex of dispersers, and the way in which the life history of one generation overlaps or tessellates with the life history of other generations in the same social group (Cant & Johnstone, 2008; Johnstone & Cant, 2010).

Figure 2.2. Human evolution.

Source: Image by pixabay.com.

Future theoretical works to further incorporate demography into models of social evolution, and measures of metapopulation fitness in wild cooperative vertebrates, offer the best hope of developing a unified "social life history theory" applicable to animal societies from insects to humans.

2.2. EARLY SOCIAL EVOLUTION

Modern man can be said to have been sent off by a two-stage speed increase in mental advancement. The first happened during the progress from a bigger arboreal primate to the primary man-gorillas (Australopithecus). Assuming the crude primate Ramapithecus is in the immediate line of parentage, as the current assessment holds, the change might have expected as much as ten million years.

Figure 2.3. Australopithecus afarensis.

Source: Image by commons.wikimedia.com.

Australopithecus was available quite a while back, and by 3,000,000 years B.P. it had speciated into a few structures, including perhaps the principal crude Homo (Tobias, 1973). As displayed the development of these halfway primates was set apart by speeding up the expansion of the mental limit. All the while, an erect stance and a stepping, bipedal movement were idealized, and the hands were shaped to gain an accurate hold (Figure 2.3).

These early men without a doubt utilized instruments to a lot more prominent degree than do modem chimpanzees. Unrefined stone executes were made by chipping, and shakes were arranged to shape what give off an impression of being the groundworks of havens.

The second, considerably quicker period of speed increase started quite a while back. It comprised basically of social advancement and probably been generally phenotypic, expanding upon the hereditary potential in the cerebrum that had gathered over the past large number of years.

The mind had arrived at a limit, and a completely new, gigantically more fast type of mental development dominated. This subsequent stage was in no sense arranged, and its true capacity is just now being uncovered.

The investigation of man's starting points can allude to two inquiries that compare the double phases of mental development:

- What elements of the climate made the primates adjust uniquely in contrast to different primates and began them along their exceptional transformative way?
- Once began, for what reason did the primates go up to this point?

The quest for the central players of early human advancement has reached over 25 years. Members in the pursuit have included Dart (1949, 1956), Bartholomew, and Birdsell (1953), Etkin (1954), Washburn, and Avis (1958), Washburn et al. (1961), Rabb et al. (1967), Reynolds (1968), Schaller, and Lowther (1969), C. J. Sprightly (1970), and Kortlandt (1972).

These scholars have focused on two undeniably significant realities concerning the science of Australopithecus and early Homo. To start with, the proof area of strength is Australopithecus africanus, the species probably going to have been the immediate progenitor of Homo, living on the open savanna.

The wear example of sand grains taken from the Sterkfontein fossils proposes a dry environment, while the pigs, pronghorns, and different warm-blooded creatures found in a relationship with the primates are of the sort generally specific for presence in meadows. The australopithecine lifestyle came as the aftereffect of a significant territory shift. The familial Ramapithecus or a significantly more predecessor structure lived in woods and was adjusted for movement through trees by arm swinging. Just a not very many other huge bodied primates have had the option to join man in passing on the woods to burn through a large portion of their lives on the ground in open natural surroundings (Figure 2.4).

Figure 2.4. Ramapithecus fossils.

Source: Image by commons.wikimedia.org.

It is not necessarily the case that groups of Australopithecus africanus burned through every one of their lives running about in the open. Some of them could have conveyed their game into caverns and, surprisingly, lived there in a long-lasting home, albeit the proof highlighting this frequently cited quality is still distant from definitive (Kurten, 1972).

Different groups might have withdrawn around evening time to the insurance of forests of trees, in the way of modem mandrills. The significant point is that much or all of the scrounging was led on the savanna. The second unconventional element of the environment of early men was the level of their reliance on creature food, obviously far more prominent than in any of the living monkeys and chimps.

The Australopithecus were catholic in their decision of little creatures. Their locales contain the remaining parts of turtles, reptiles, snakes, mice, hares, porcupines, and other little, weak prey that probably proliferated on the savanna. The man-chimps likewise chased mandrills with clubs. From an examination of 58 mandrill skulls, Dart assessed that all had been carried somewhere near hits to the head, 50 from the front and the rest behind.

The Australopithecus likewise seem to have butchered bigger creatures, including the goliath Siva there, or homed giraffes, and dinotheres, elephantlike structures with tusks that bent descending from the lower jaws.

In early Acheulean times, when Homo erectus started utilizing stone tomahawks, a portion of the types of enormous African vertebrates became wiped out. It is sensible to assume that this impoverishment was because of unreasonable predation by the inexorably skillful groups of men (Martin, 1966).

2.3. LATER SOCIAL EVOLUTION

Autocatalytic responses in living frameworks never extend to limitlessness. Organic boundaries typically change in a rate-subordinate way to slow development and ultimately stop it. Yet, supernaturally, this has not yet occurred in human advancement.

The expansion in mind size and the refinement of stone antiquities show a steady improvement in intellectual ability all through the Pleistocene. With the presence of the Mousterian instrument culture of Homo sapiens neanderthalensis about quite a while back, the pattern built up speed, giving way in Europe to the Upper Paleolithic culture of Homo s. sapiens around 40,000 years B.P.

Beginning around quite a while back, horticulture was designed and spread, populaces expanded immensely in thickness, and the crude agrarian groups gave way locally to the persistent development of clans, chiefdoms, and states. At last, after A.D. 1400 European-based human advancement changed gears once more, and information and innovation became dramatically as well as super exponentially.

There is not a great explanation to trust that during this last run there has been an end in the development of either intellectual ability or the inclination toward exceptional social ways of behaving. The hypothesis of population hereditary qualities and investigations on different organic entities demonstrate the way that significant changes can happen in the range of under 100 ages, which for man arrives at back just to the hour of the Roman Empire.

2,000 ages, generally the period since commonplace Homo sapiens attacked Europe, is sufficient opportunity to make new species and to form them in significant ways. Despite the fact that we don't have the foggiest idea how much mental advancement has really happened, it would be bogus to expect that modem civic establishments have been fabricated totally on capital gathered during the long stretch of the Pleistocene.

Since hereditary and social global positioning frameworks work on equal tracks, we can sidestep their differentiation for the second and return to the subject of the main players in later human social advancement in its broadest sense. Seed eating is a conceivable clarification to represent the development of primates onto the savanna, and the shift to major game hunting could represent their development to the Homo erectus grade (Figure 2.5).

Figure 2.5. Homo sapiens.

Source: Image by commons.wikimedia.com.

Be that as it may, was the variation to bunch predation enough to convey development the entire way to the Homo sapiens grade and farther, to horticulture and human advancement? Anthropologists and scholars don't consider the impulse to have been adequate. They have pushed the accompanying series of extra factors, which can act separately or in blend.

2.4. THE MULTIPLIER EFFECT IN SOCIAL EVOLUTION

Social organization is the class of phenotypes furthest removed from the genes. It is derived jointly from the behavior of individuals and the demographic properties of the population, both of which are themselves highly synthetic properties. A small evolutionary change in the behavior pattern of individuals can be amplified into a major social effect by the expanding upward distribution of the change into multiple facets of social life.

This phenomenon can be referred to as the multiplier effect. Consider, for instance, the varying social associations of the connected olive monkey (*Papio anubis*) and hamadryas primate (P. hamadryas). These two species are so close hereditarily that they interbreed broadly where their reaches cross-over and could sensibly be named something like subspecies.

The hamadryas male is recognized by its exclusive mentality toward females, which is aggregate and extremely durable, though the olive male endeavors to suitable females just around the hour of their estrus. This obviously hereditary contrast is only one of degree, and would barely be perceptible if one' s interest were confined in every species to the exercises of a solitary prevailing male and one partner female.

However, this attribute alone is sufficient to represent significant contrasts in friendly design, influencing the size of the soldiers, the relationship of troops to each other, and the relationship of guys inside each troop (Kummer, 1971). Multiplier impacts can speed social advancement even more when a singular's way of behaving is unequivocally affected by the particularities of its social experience.

This cycle, called socialization, turns out to be progressively unmistakable as one maneuvers up phylogenetically into additional shrewd species, and it arrives at its most extreme impact in the higher primates. Albeit the proof is still generally inferential, socialization seems to enhance phenotypic contrasts among primate species.

For instance, take the wandering pathways of improvement of social conduct saw in youthful olive mandrills (*Papio anubis*) instead of Nilgiri langurs (Presby tis johnii) (Eimerl and DeVore, 1965; Poirier, 1972b). The primate newborn child remains nearby its mom during the main month of its life, and the mother beats the methodology of different females down. In any case, a short time later the developing newborn child relates uninhibitedly with grown-ups.

It is even moved toward by the guys, who frequently gravitate toward to the mother, licking their lips in the run of the mill placating sign to be close to the young person. From the age of nine months, male monkeys dynamically lose the security of their moms, who reject them with expanding seriousness. Thus, they come to blend with different individuals from the troop considerably more rapidly and openly.

The social design of the olive monkey is reliable with this program of socialization. Grown-up guys and females blend uninhibitedly, and fringe gatherings of guys and singular people are uncommon or missing. Langur social improvement is definitely more sex arranged than that of monkeys.

The baby is surrendered promptly to other grown-up females, who pass it around. In any case, it has little contact with grown-up guys, who are pursued away at whatever point they upset the young person. Adolescent guys start to connect with grown-up guys solely after eight months, while youthful females don't allow contact until the beginning of sexual movement at three years old years.

The youthful guys invest the majority of their free energy playing. As the playfighting becomes harsher and requires more space, they will generally float to the fringe of the gathering, well separated from the newborn children and grown-ups.

The langur society mirrors this type of isolated raising. Grown-up guys and females will more often than not stay separated. Gatherings of fringe guys are normal. what's more, they frequently collaborate forcefully with the prevailing guys inside the soldiers trying to infiltrate and acquire command.

2.5. THE EVOLUTIONARY PACEMAKER AND SOCIAL DRIFT

The multiplier effect, whether absolutely hereditary in premise or built up by socialization and different types of learning, makes conduct the piece of the aggregate probably going to change because of long haul changes in the climate. It follows that when development includes both construction and conduct ought to change first and afterward structure.

As such, conduct ought to be the developmental pacemaker. This is an old thought, with roots reaching out back similar to the 6th version of Darwin's Origin of Species (1872) and the guideline of Funktionswechsel communicated by Anton Dohm (1875).

Dohm hypothesized that the capacity of an organ, which by and large we can view to be most plainly communicated in its way of behaving, is constantly changing and dichotomizing over numerous ages as per the experience of the creature. Changes in the construction of the organ address facilities to these useful movements.

Among late zoologists, Wickler (1967a,b) has most explicit contended a similar perspective regarding conduct, referring to numerous models from birds and fishes. Among the tetraodontiform fishes, to take one of the less difficult and more clear cases, various species can swell themselves colossally with water or air as a defensive gadget against hunters.

In youthful porcupine fishes of the class Diodon, the middle blades vanish into pockets of the skin that overlay internal during expansion. The swelled stage has become irreversible in the diodontid family Hyosphaera, while the tetraodontid globe fish, Kanduka michiei, not exclusively is for all time expanded yet in addition has lost the dorsal balance and diminished the butt-centric blade to minimal structure. Social way of behaving likewise much of the time fills in as a transformative pacemaker. The whole course of ritualization, during which a way of behaving is changed by advancement into a more effective flagging gadget, regularly includes a conduct change followed by morphological adjustments that improve the perceivability and peculiarity of the way of behaving.

The overall lability of conduct drives unavoidably to social float, the arbitrary difference in the way of behaving and method of association of social orders or gatherings of social orders. The term arbitrary implies that the conduct distinctions are not the consequence of variation to the specific circumstances by which the living spaces of one society vary from those of different social orders.

Assuming that the difference has a hereditary premise, the innate part of social float is essentially equivalent to hereditary float, a developmental peculiarity. How much difference inside a populace of social orders is the amount of the fluctuations because of hereditary float, custom float, and their cooperation.

2.6. THE DUALITIES OF EVOLUTIONARY BIOLOGY

The theories of behavioral biology are riddled with semantic ambiguity. Like buildings constructed hastily on unknown ground, they sink, crack, and fall to pieces at a distressing rate for reasons seldom understood by their architects. In the special case of sociobiology, the unknown substratum is usually evolutionary theory (Figure 2.6).

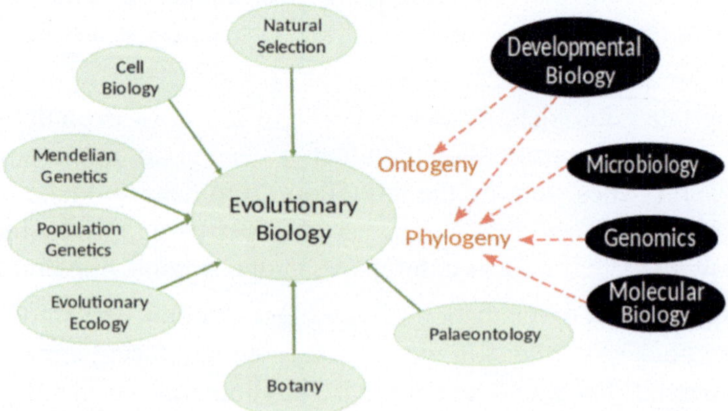

Figure 2.6. Evolutionary biology branches.

Source: Image by commons.wikimedia.com.

We should therefore set out to map the soft areas in the relevant parts of evolutionary biology. With remarkable consistency the most troublesome evolutionary concepts can be segregated into a series of dualities. Some are simple two-part classifications, but others reflect more profound differences in levels of selection and between genetic and physiological processes.

2.6.1. Adaptive Versus Nonadaptive Traits

A characteristic can be supposed to be versatile assuming it is kept up with in a populace by choice. We can put the matter all the more unequivocally by saying that another quality is nonadaptive, or "strange," assuming it decreases the wellness of people that reliably manifest it under ecological conditions that are regular for the species.

As such, degenerate reactions in strange conditions may not be nonadaptive-they may basically reflect adaptability in a reaction that is very versatile in the conditions conventionally experienced by the species. A characteristic can be changed from a versatile to a nonadaptive status by a basic change in the climate.

For instance, the sickle-cell quality, not set in stone by the heterozygous condition of a solitary quality, is versatile under day-to-day environments in Africa, where it gives a level of protection from falciparum jungle fever. In Americans of African drop, it is nonadaptive, for the basic explanation that its conveyors are not generally faced by intestinal sickness.

The inescapable job of regular choice in molding all classes of characteristics in creatures can be decently called the focal authoritative opinion of developmental science. Whenever persistently squeezed, this suggestion may not create an essential fact of the matter, but rather it is, as G. C. Williams disarmingly put the matter, the light and the way.

A huge piece of the commitment of Konrad Lorenz and his kindred ethologists can be outlined in a similar representation. They persuaded us that way of behaving and social construction, similar to any remaining natural peculiarities, can be concentrated as "organs," augmentations of the qualities that exist distract of their predominant versatile worth.

Monadaptive Versus Poly-adaptive Traits

Social evolution is set apart by areas of strength for rehashed of broadly separate phylogenetic gatherings. The disarray intrinsic in this situation is demolished by the still coarse and moving nature of our nomenclatural frameworks. Preferably, we ought to attempt to have a term for each major utilitarian class of social way of behaving.

This semantic refinement would bring about most sorts of social way of behaving being perceived as monadaptive, that is to say, having just a single capacity. In our clearly flawed language, nonetheless, most ways of behaving are falsely understood as polyadaptive.

Think about the polyadaptive idea of II animosity, or "II agonistic way of behaving," in monkeys. Guys of langurs, patas, and numerous different species use animosity to keep troop separation. Comparative way of behaving is additionally utilized by a variety of animal categories, including langurs, to lay out and to support pecking orders.

Male hamadryas mandrills use animosity to group females and deter them from leaving the arrays of mistresses. Hostility, to put it plainly, is an unclear term used to assign a variety of ways of behaving, with different capacities, that we naturally feel look like human animosity.

Some friendly ways of behaving by the by remain genuinely polyadaptive even after they have been semantically filtered. Allogrooming in rhesus monkeys, for models, serves the normally higher primate capacity of pacification and bond upkeep.

However, it holds a second, evidently cruder, purging capacity, since monkeys kept in seclusion frequently foster serious invasions of lice. In some bird species, running way of behaving without a doubt serves the double capacity of hunter avoidance and improvement in scavenging effectiveness.

2.6.2. Reinforcing Versus Counteracting Selection

A single force in natural selection acts on one or more levels in an ascending hierarchy of units: the individual, the family, the group, and possibly even the entire population or species. If affected genes are uniformly favored or disfavored at more than one level, the selection is said to be reinforcing.

Evolution, meaning changes in gene frequency, will be accelerated by the additive effects inserted at multiple levels. This process should offer no greater problem to mathematicians. By contrast, the selection might be counteracting in nature: genes favored by a selective process at the individual level could be opposed by the same process at the family level, only to be favored again at the population level, and so on in various combinations. The compromise gene frequency is of general importance to the theory of social evolution, but it is mathematically difficult to predict.

2.6.3. Ultimate versus Proximate Causation

The division between functional and evolutionary biology is never more clearly defined than when the proponents of each try to make a pithy statement about causation. Consider the problem of aging and senescence. Contemporary functional biologist is preoccupied with four contending speculations of maturing, all stringently physiological: pace of-living, collagen wear, autoimmunity, and substantial change (Curtis, 1971).

On the off chance that at least one of these elements can be solidly ensnared such that records for the entire cycle in the existence of an individual, the more barely prepared organic chemist will consider the issue of causation tackled. Be that as it may, just the general causation will have been illustrated.

In the interim, like dwelling in another land, the hypothetical populace geneticist chips away at senescence as an interaction that is shaped in time in order to amplify the regenerative wellness specifically conditions (Hamilton, 1966).

These experts know about the presence of physiological cycles yet see them dynamically as components to be jiggered to get the ideal season of senescence as per the timetables of survivorship and richness that win in their hypothetical populaces. This approach endeavors to take care of the issue of extreme causation.

2.6.4. Ideal versus Optimum Permissible Traits

When organisms are thought of as machine analogs, their evolution can be viewed as a gradual perfecting of design. In this conception there exist ideal traits for survival in particular environments. There would be the ideal hammer bill and extrusible tongue for woodpeckers, the ideal caste system for army ants, and so forth.

But we know that such traits vary greatly from species to species, even those belonging to the same phyletic group and occupying the same narrowly defined niche. In particular, it is disconcerting to find frequent cases of species with an advanced state or intermediate states of the same character.

Specifically, finding continuous instances of species with a high-level state or middle conditions of a similar character is unsettling. Take the hypothetical issue made by the crudely friendly bug species. Why have they advanced no further?

Two outrageous conceivable outcomes can be imagined (Wilson, 1971a). In the first place, there may be named the "disequilibrium case." This implies that the species is still effectively developing toward a higher social level. The circumstance can emerge assuming social advancement is delayed to such an extent that the species is set out on a specific versatile course yet is still on the way.

2.6.5. Deep versus Shallow Convergence

At this phase of our insight, it is attractive to start an examination of developmental union as such, for the explanation that a similarity perceived between two ways of behaving in one case might be a considerably more significant and huge peculiarity than a relationship perceived for another situation.

It will be helpful to make an unpleasant differentiation between occurrences of transformative combinations that are profound and those that are shallow. The essential main attributes of a profound combination are two: the intricacy of the variation and the degree to which the species has coordinated its lifestyle around it.

The eye of the vertebrate and the eye of the cephalopod mollusk comprise a natural illustration of an extremely profound assembly. Different qualities related with profound union, however not basically characterizing the peculiarity, are level of distance in phylogenetic beginning, which decides the added measure of development the two phyletic lines should venture out to the mark of assembly, and strength.

Extremely shallow union is much of the time set apart by hereditary lability. Related species, and some of the time populaces inside similar species, vary in how much they show the quality, and some don't have it by any means.

2.6.6. Grades versus Clades

Evolution consists of two simultaneously occurring processes: while all species are evolving vertically through time, some of them split into two or more independently evolving lines. In the course of vertical evolution, a species, or a group of species, ultimately passes through a series of stages in certain morphological, physiological, or behavioral traits (Figure 2.7).

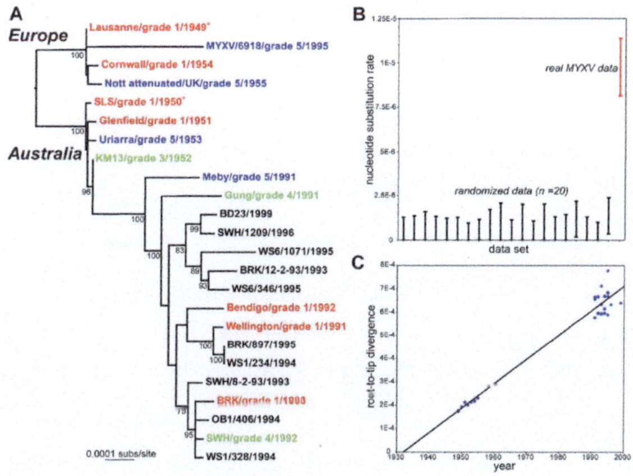

Figure 2.7. Evolutionary history of MYX.A.

Source: Image by fickr.com.

If the stages are distinct enough, they are referred to as evolutionary grades. Phylogenetically remote lines can reach and pass through the same grades, in which case we speak of the species making up these lines as being convergent with respect to the trait.

A different developing line is alluded to as a clade, and a spreading outline that shows how species split and structure new species is known as a cladogram (Simpson, 1961; Mayr, 1969). The full phylogenetic tree contains the data of the cladogram, in addition to some proportion of how much dissimilarity between the branches, plotted against a period scale.

Sociobiologists are keen on both the transformative grades of social way of behaving and the phylogenetic connections of the species inside them. A phenomenal worldview from the writing of social wasps.

2.6.7. Instinct versus Learned Behavior

Throughout the entire existence of science, no differentiation has delivered a more prominent semantic quagmire than the one among sense and learning. A few ongoing essayists have endeavored to avoid the problem out and out by announcing it a nonproblem and declining to proceed with the sense learning polarity as a component of modem language.

As a matter of fact, the differentiation stays a valuable one, and the semantic trouble can be cleared up rather without any problem. The way in to the issue is the acknowledgment that sense, or intrinsic way of behaving, as it is frequently called, has been naturally characterized in two altogether different ways:

- An intrinsic social distinction between two people or two species is one that is put together to some degree to a limited extent with respect to a hereditary contrast. We then, at that point, talk about contrasts in the genetic part of the way of behaving, or of inborn contrasts in conduct, or, most freely, of contrasts in impulse.
- A nature, or inborn way of behaving, is a way of behaving that either is dependent upon somewhat little alteration in the lifetime of the organic entity, or changes next to no all through the populace, or (ideally) both.

The principal definition can be made exact, since it is only an exceptional instance of the typical differentiation made by geneticists among acquired and earth forced variety. It requires, nonetheless, that we distinguish a distinction between at least two people.

Consequently, by the principal definition, blue eye variety in individuals can be ended up being hereditarily unique in relation to brown eye tone. Yet, it is useless to find out if blue eye tone not set in stone by heredity or climate. Clearly, both the qualities for blue eye tone and the climate added to the end result.

The main valuable inquiry regarding the primary definition is whether individuals that foster blue eye tone rather than earthy colored eye tone do as such partially in light of the fact that they have qualities not the same as those that control earthy colored eye tone. A similar thinking can be stretched out without change to various examples of social way of behaving.

The second natural meaning of sense can be generally promptly gotten a handle on by thinking about one of the outrageous models that fits it. The guys of moth species are typically drawn in just to the sex pheromones radiated by the females of their own species. Sometimes they might be "tricked" by the pheromones of other, firmly related species, however seldom to the degree of finishing fornication.

The sex pheromone of the silkworm moth (*Bambyx mari*) is 10, 12-hexadecadienol. The male answers just to this substance, and he is more delicate by a few significant degrees to one specific mathematical isomer (trans10-cis-12-hexadecadienol) than to different isomers.

Additionally, the separation happens at the level of the sensilla trichodea, the hairlike olfactory receptors circulated over the radio wires. Just when these organs experience the right pheromone do they send anxious motivations to the mind, setting off the efferent progression of orders that start the sexual reaction.

In not the remotest sense is learning associated with such a machinelike reaction, which is typical of a significant part of the way of behaving of arthropods and different spineless creatures. Not many invertebrate zoologists have a reluctant outlook on implying this way of behaving as natural or intuitive, and they have as a primary concern both the first and second definitions.

At the contrary limit, we have the plastic characteristics of human discourse and vertebrate social association, and nobody feels right in naming these qualities instinctively continuously. A second's appearance on the transitional cases uncover that they can't be ordered by a severe measure equivalent to the presence or nonappearance of a hereditary part utilized in the primary definition.

Thus, the subsequent definition can never be exact, and it truly has educational substance just when applied to the outrageous cases.

2.7. THE REVERSIBILITY OF SOCIAL EVOLUTION

Two broad generalizations have begun to emerge that will be reinforced in subsequent chapters: the ultimate dependence of particular cases of social evolution on one or a relatively few idiosyncratic environmental factors; and the existence of antisocial factors that also occur in a limited, unpredictable manner (Figure 2.8).

Figure 2.8. Reversibility of human evolution.

Source: Image by piqsel.com.

In the event that the antisocial compels come to win eventually after friendly development has been started, it is hypothetically workable for social species to be gotten back to a lower social state or even to the singular condition. Something like two such cases have been proposed. Michener (1964b, 1965) saw that allodapine honey bees of the family Exoneurella are somewhat less than completely friendly, since the females scatter from the home prior to being joined by their girls. This condition seems to have been gotten from the conduct actually shown by the firmly related variety Exoneura, in which the mother and little girls stay in affiliation.

Michener (1969) additionally noticed that inversions might have happened in the crudely friendly types of the halictine sweet honey bees. The most probable particular power, gathered from field concentrates on the halictines, is the unwinding of tension from such home parasites as mutillid wasps. The subsequent case is from the vertebrates.

In the ploceidae weaver birds, as in most other passerine gatherings, the species that home in timberlands and feed fundamentally on bugs are single in propensity, or at generally regional. As indicated by Crook (1964), these species have advanced from other species that live in savannas, eat seeds-and, in the same way as other passerine gatherings comparatively specific home in pilgrim gatherings, in a couple of instances of exceptionally huge size.

2.8. SOCIAL EVOLUTION AND PARENTAL CARE IN THE INSECTS AND MAMMALS

One of a handful of the truly central contrasts among bug and vertebrate social orders exists in the domain of parental consideration. The types of social bugs show variety in the structure and force of consideration paid the youthful, and this variety is just feebly corresponded with the level of intricacy of social association.

Figure 2.9. Parental care in monkeys.

Source: Image by piqsels.com.

In the absolute most exceptional species practically no contact happens between the "adults and youthful structures. Additionally, likely parent-posterity cooperation affects the improvement of social way of behaving. For instance, the stingless honey bees (Meliponini) seal their eggs in waxen brood chambers and have no communication with the creating youthful until they arise as full grown-ups, all set to work (Figure 2.9).

Despite this situation, the stingless honey bees have the absolute most complex social orders tracked down in every one of the bugs. Vertebrate species, conversely, show areas of strength for a between how much parental consideration and the intricacy of social association. Moreover, the way of behaving of the parent unequivocally impacts the social advancement of the posterity.

Both of these relations are very much set apart inside the well evolved creatures, particularly the primates. To show the more precise connection that exists between brood care and social association in the well evolved creatures, we can do no better compared to look at the limits inside the primates.

The tree wenches, comprising the family Tupaiidae, are crude to the point that their arrangement as primates has been questioned, a few creators liking to put them inside the Insectivora near the elephant vixens. In any case, the heaviness of proof, mostly osteological and serological in nature, demonstrates that they fit near while possibly not straightforwardly in accordance with the beginning of the primates.

Tree wenches turn out to have a basic however very curious type of parental consideration, portrayed by non-attendance with respect to the mother. Martin (1968) had the option to follow the total rearing cycle in six sets of Tupaia Gus Belangeri under seminatural conditions. Social association is negligible.

Sexual pair holding areas of strength for is, the male imprints both the enclosure and his mate with fragrance. The male and female rest together in a home that is developed for the most part by the male. Whenever the female becomes pregnant she constructs a subsequent home, in which she later bears her young.

The litter size is regularly two. The mother before long passes on the defenseless babies and gets back to the primary home to rejoin the male. From that point, she visits the nursery just like clockwork. The babies move starting with one of her six areolas then onto the next in an unsystematic manner.

Following a couple of moments, the mother shakes them off and takes off. The youthful are passed on to prepare themselves and, probably, to clear out their own dung. The grown-ups don't recover them when they are left external the home, and the newborn children don't articulate misery calls.

Guardians compelled to remain excessively near their young frequently kill and eat them. At the point when they are around 33 days old, the youthful tree vixens begin rising up out of their home for short rummaging trips, and the musicality of maternal visits starts steadily to separate.

At first the youthful re-visitation of their own home around evening time and at whatever point they are scared, yet following 3 days they shift to the parental home. Sexual development is reached at around 90 days old enough. A short time later the youthful obviously disperse to track down mates and regions of their own.

Among the nonhuman primates, chimpanzees are at the contrary limit from the tree wenches in both parental consideration and social association. The improvement of the youthful chimpanzee has been concentrated on in hostage bunches by numerous examiners, however our head comprehension of the cycle comes from the naturalistic investigations directed by Jane van Lawick-Goodall (1967; 1968a,b; 1971) on the wild populace of the Gombe Stream National Park in Tanzania.

The principal meaning of her exploration has been to show how incredibly unobtrusive, convoluted, and, surprisingly, human is the social improvement of the youthful chimpanzee. The interaction unfurls over a time of somewhat more than decade.

It comprises basically of the sluggish acquiring of locomotory capability, during which the baby passes on the mother for protracting spans to investigate the climate, control articles, and play with different individuals from the troop. At first the young person is comprehensively endured by the grown-ups it experiences, and their well-disposed reactions contribute a significant piece of its socialization.

Yet, as it approaches development, it starts to get repels from grown-ups. Forceful associations strengthen as the chimpanzee next tracks down its direction into the grown-up progressive system. Albeit the mother gently dismisses her young person' s endeavors to nurture during the weaning time frame, she stays a partner and comfort all through its youth.

2.9. THE FOUR PINNACLES OF SOCIAL EVOLUTION

To visualize the main features of social behavior in all organisms at once, from colonial jellyfish to man, is to encounter a paradox. We should first note that social systems have originated repeatedly in one major group of organisms after another, achieving widely different degrees of specialization and complexity.

Four groups occupy pinnacles high above the others: the colonial invertebrates, the social insects, the nonhuman mammals, and man. Each has basic qualities of social life unique to itself. Here, then, is the paradox.

Although the sequence just given proceeds from unquestionably more primitive and older forms of life to more advanced and recent ones, the key properties of social existence, including cohesiveness, altruism, and cooperativeness, decline.

It seems that social evolution slowed as the body plan of the individual organism became more elaborate. The colonial invertebrates, including the corals, the jelly fish like siphonophores, and the bryozoans, have come close to producing perfect societies.

The singular individuals, or zooids as they are called, are by and large completely subjected to the state in general in work as well as more in a real sense, through close and completely associated actual association. So outrageous is the specialization of the individuals, thus exhaustive their get together into actual wholes, that the province can similarly well be called an organic entity.

It is feasible to exhibit the types of provincial spineless creatures into a subtly reviewed transformative series from those shaping bunches of generally free and independent zooids to others with settlements that are practically indistinct from multicellular living beings.

The higher social bugs, made out of the subterranean insects, termites, and certain wasps and honey bees, structure social orders that are substantially less than awesome. Certainly, they are portrayed by sterile standings that are benevolent in the help of the mother sovereign. Additionally, selfless way of behaving is unmistakable and changed.

It incorporates the re gurgitation of stomach items to hungry nestmates, self-destructive weapons, for example, separable stings and detonating midsections utilized with regards to the state, and other specific reactions. The positions are genuinely changed to fill specific roles and are bound to each other by close, complicated types of correspondence.

Besides, people can't live separated from the state for more than brief periods. They can perceive standings however not individual nestmates. In a word, the bug society depends on unoriginal closeness. In any case, these similitudes to the settlements of lower spineless creatures are adjusted by a few intriguing characteristics of autonomy.

Social bugs are genuinely independent substances. The mystery of their prosperity is truth be told the capacity of a state to dispatch independently versatile foragers that return occasionally to the headquarters. Likewise, the sovereigns are not generally the elite egg layers. Female specialists once in a while embed eggs of their own into the blood cells.

Since these eggs are unfertilized, they form into guys. The proof areas of strength for are in certain types of subterranean insects, honey bees, and wasps a low-keyed battle consistently happens among sovereigns and laborers for the chance to deliver children. Struggle is once in a while plain in additional crude structures.

Gatherings of female wasps beginning a state together fight for predominance and the egg-laying freedoms that go with the alpha position. Failures proceed as laborers, now, and again covertly embedding eggs of their own into void brood cells.

Additionally, honey bee sovereigns control their girls by animosity, going after them at whatever point they endeavor to lay eggs. Assuming that the sovereign is taken out from the somewhat basic wasp and honey bee social orders, sure of the laborers battle among themselves for the option to supplant her.

Forcefulness and disagreement are conveyed a lot further in vertebrate social orders, including those of well evolved creatures. Self-centeredness manages the connections between individuals. Sterile stations are obscure, and demonstrations of selflessness are rare and commonly coordinated exclusively toward posterity.

Every individual from the general public is a possibly autonomous, imitating unit. Albeit a creature's possibilities of endurance are decreased on the off chance that it is constrained into a singular presence, bunch participation isn't obligatory on an everyday premise for all intents and purposes in provincial spineless creatures and social bugs.

Every individual from a general public is all alone, taking advantage of the gathering to acquire food and asylum for itself and to raise however many posterity as could be allowed. Participation is normally simple. It addresses a concession by which individuals can raise their own endurance and conceptive rates over those that would gather from a single life.

By human norms, life in a fish school or a monkey troop is tense and merciless. The wiped out and harmed are normally left where they fall, without even a respite in the standard business of taking care of, resting, and mating. The demise of a prevailing male is generally trailed by just a change in the pecking order, maybe went with, as on account of langurs and lions, by the homicide of the pioneer's most youthful posterity.

Individuals remain basically vertebrate in their social construction. Yet, they have conveyed it to a degree of intricacy so high as to comprise an unmistakable, fourth apex of social advancement. They have broken the old vertebrate restrictions not by diminishing self-centeredness, yet rather by procuring the knowledge to counsel the past and to design the future.

Individuals lay out lengthy recollected agreements and beneficially take part in demonstrations of corresponding charitableness that can be separated throughout significant stretches of time, to be sure over ages. People instinctively bring kinfolk determination into the analytics of these connections; they are distracted with family relationship binds to a degree incomprehensible in other social species.

Their exchanges are made even more effective by a remarkable grammatical language. Human social orders approach the bug social orders in helpfulness and far surpass them in powers of correspondence. They have turned around the descending pattern in friendly advancement that beat one billion years of the past history of life.

At the point when the human type of social association is set in this viewpoint, it maybe appears to be less amazing that it has emerged just a single time, though the other three pinnacles of advancement have been scaled over and again by autonomously developing lines of creatures. Why has the general pattern been descending?

The explanation should have something to do with the more noteworthy actual flexibility of the lower spineless creatures. Since their body plan is so rudimentary, such pioneer creatures as coral polyps and bryozoans can be terribly adjusted to allow a genuine actual association with each other.

Contrasted and bugs and vertebrates, they require less "revamping" of nerve cells, rerouting of circulatory frameworks, and different changes of organ frameworks expected to organize pilgrim physiology. The for the most part stationary propensities for zooids additionally make it more straightforward for them to be combined. Notwithstanding, this benefit isn't the definitive one; probably the most intricate invertebrate provinces, including those of siphonophores and thaliaceans, are additionally the most motile. The basic body development of lower spineless creatures likewise makes it feasible for them to repeat by straightforwardly sprouting off new people from old.

States in this manner comprise of hereditarily indistinguishable people. Furthermore, this, in the last examination, is the main component of all. Outright hereditary personality makes conceivable the advancement of limitless benevolence.

It is now the reason for the outrageous specialization and coordination of substantial cells and organs inside the metazoan body. The most developed pioneer spineless creatures have followed basically similar street, prompting superorganisms whose organs are made by the outrageous adjustment of zoo ids.

The social bugs have none of the preadaptation of the lower spineless creatures. Their bodies are in numerous ways as intricately built as those of the vertebrates, and they are completely motile. Actual association is incomprehensible.

However, they have created unselfish ranks and levels of province mix nearly however outrageous as it very well might be feasible to envision. I might want to propose that piece of the clarification of this accomplishment is the sheer tremendousness of the example size. North of 800,000 types of bugs have been depicted, comprising more than 3/4 of the relative multitude of known sorts of creatures on the planet.

In the late Paleozoic Era, the predecessors of the living bugs were among the principal trespassers of the land, and they made the most of this major biological open door. While the sea and new water were at that point packed with the agents of numerous creature phyla, a large portion of which had begun in Precambrian times, the land resembled another planet, loading up with vegetation and almost absent any trace of creature contenders.

The outcome was a versatile radiation of species unrivaled previously or since. The absolutely measurable contention expresses that out of this colossal exhibit of new sorts, it was more plausible for essentially a couple of outrageous social structures to emerge in correlation, say, with the simple 7,000 types of annelid worms or 5,300 types of starfish and different echinoderms.

The contention can maybe be made clearer by envisioning a substantial model: in the event that the pace of development of cutting-edge sociality were 10–12 for every species each year for creatures for the most part, 800,000 bug species would surely have accomplished it ordinarily by chance alone, while 10,000 species having a place with another phylum could never do so.

At last, the regularly vertebrate characteristics of further developed correspondence, individual acknowledgment, and expanded conduct adjustment make conceivably still one more property critical: the arrangement of self-centered subgroups inside the general public.

It is workable for mated matches, parent-posterity gatherings, bunches of kin and other close family, and even clubs of irrelevant people to exist inside social orders without losing their different characters.

Each seeks after its own finishes, forcing extreme cutoff points on how much the general public overall can work as a unit. The regular vertebrate society, to put it plainly, favors individual and in-bunch endurance to the detriment of cultural respectability.

2.10. CONCLUSION

In the conclusion of the chapter, it discussed about the early social evolution and the later social evolution. It also provides highlights on the significance of multiplier effect in social evolution. In this chapter, the evolutionary pacemaker and social drift have also been discussed.

Towards the end of the chapter, several dualities of evolutionary biology have also been discussed. It also addresses the concept of reversibility of social evolution. This chapter addresses the social evolution and parental care in the insects and the mammals. In this chapter, the four pinnacles of social evolution have also been discussed.

REFERENCES

1. (n.d). *Evolution of Human Society ; Nature and Characteristics*. [e-Book] https://egyankosh.ac.in. Available at: https://egyankosh.ac.in/bitstream/123456789/17114/1/Unit-2.pdf (accessed on 27 June 2022).
2. (n.d). *Social Evolution and Societal Change*. [e-Book] http://web.stanford.edu/. Available at: http://web.stanford.edu/~mshanks/MichaelShanks/files/134368.pdf (accessed on 27 June 2022).
3. Di Nuoscio, E., (2001). Social evolution, sociology of. *International Encyclopedia of the Social & Behavioral Sciences*, 14286–14290.
4. Klaus, E., (1992). *Contradictions and Social Evolution: A Theory of the Social Evolution of Modernity*. [e-Book] https://www.ssoar.info/. Available at: https://www.ssoar.info/ssoar/bitstream/handle/document/1520/ssoar-1992-eder-contradictions_and_social_evolution.pdf?sequence=1 (accessed on 27 June 2022).
5. Rubenstein, D., & Abbot, P., (2017). *The Evolution of Social Evolution*. [e-Book] https://www.cambridge.org/. Available at: http://www.columbia.edu/~dr2497/download/rubenstein-and-abbot-2017a-book.pdf (accessed on 27 June 2022).
6. Wilson, E., (1975). *Sociobiology the Abridged Edition*. [e-Book] The Belknap Press of Harvard University Press. Available at: https://teoriaevolutiva.files.wordpress.com/2014/02/edward_o-_wilson_sociobiology_the_abridged_editbookos-org.pdf (accessed on 27 June 2022).
7. Wilson, E., (2000). *Sociobiology the New Synthesis, Twenty-Fifth Anniversary Edition*. [e-Book] Belknap Press. Available at: https://www.hup.harvard.edu/catalog.php?isbn=9780674002357&content=toc (accessed on 27 June 2022).

CHAPTER 3

Population Biology Principles

CONTENTS

3.1. INTRODUCTION

In a particular geographic area at the same time, with the capability of interbreeding the number of organisms of the same species that live is known as population. Individuals must be able to mate with any other member of a population and produce fertile offspring for interbreeding to occur.

Not all individuals are equally able to survive and reproduce and however, populations contain genetic variation within themselves. On various different scales populations can occur. To a spatially small area, i.e., the fish in one pond a local population can be confined.

On a regional, countrywide, island or continental scale; it may even make up the entire species, however, this locality can operate. It is called a metapopulation if individuals of local populations are able to disperse between other local populations.

The study of population characteristics and the factors that affect their size and distribution is known as population biology. The study of how these populations interact with the environment is called population ecology.

3.2. POPULATION BIOLOGY

The study of the factors determining the size and distribution of a population, as well as the ways in which populations change over time is known as population biology. When researchers merged aspects of population ecology with aspects of population genetics, the discipline of population biology dates back to the 1960s.

The measurement of physical (abiotic) factors and the living (biotic) factors that may affect population numbers and it employs a traditional empirical approach which consists of observation of the numbers of individuals in a population and the variation in those numbers over time and space.

Exponential growth is given optimum conditions, the populations of most organisms grow at a constant rate of increase, doubling in size at regular intervals. Such as disease, predation, or harsh climates are exponential population growth which is explosive but it is usually opposed by factors that reduce numbers (Figure 3.1).

Where the rapidly growing population slows and reaches a stable but dynamic equilibrium at or near the carrying capacity of the environment, the result is logistic growth. Such as migratory locusts show boom and bust cycles are the populations of some species: explosive growth produces vast numbers of individuals that eventually overwhelm the carrying capacity and a catastrophic dieback follows, in these cases.

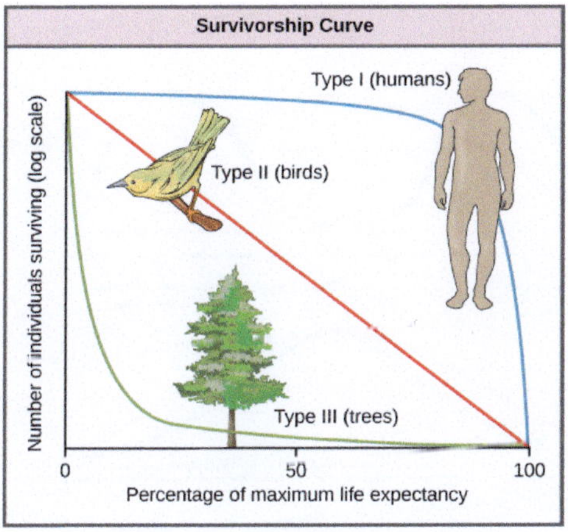

Figure 3.1. Survivorship curve.

Source: Image by commons.wikimedia.com.

Within the population, the rate of growth of a population is the net result of the gains and losses from a number of intrinsic factors operating. These include fecundity, natality, longevity, mortality, immigration, emigration, and life span. By these interacting factors create, patterns of survivorship and age structure indicate what general role a species plays in the ecosystem and show how a population is growing.

Including the numbers of competitors both interspecific and intraspecific, density-dependent biotic factors that decrease natality or increase mortality as well as predators, parasites, prey, and other interactive species. Through excessive intraspecific competition for limited resources, stress, and overcrowding are other density-dependent biotic factors that limit population size.

By physical factors the local distribution pattern of populations of most species is limited such as, moisture, light, pH, temperature, salinity, and soil quality. Within their areas of distribution, animals occur in varying dispersal patterns (either spaced evenly or clumped into herds) and in varying densities (either scattered thinly or crowded) (Figure 3.2).

In ecology and management, population density and dispersion are often studied together, and are important. By direct visual count, and by trapping, density is measured using pelt records, monitoring vocalization frequencies, collecting fecal pellets, and so on. How a population is age-structured can be shown by these life tables constructed from these data show precisely.

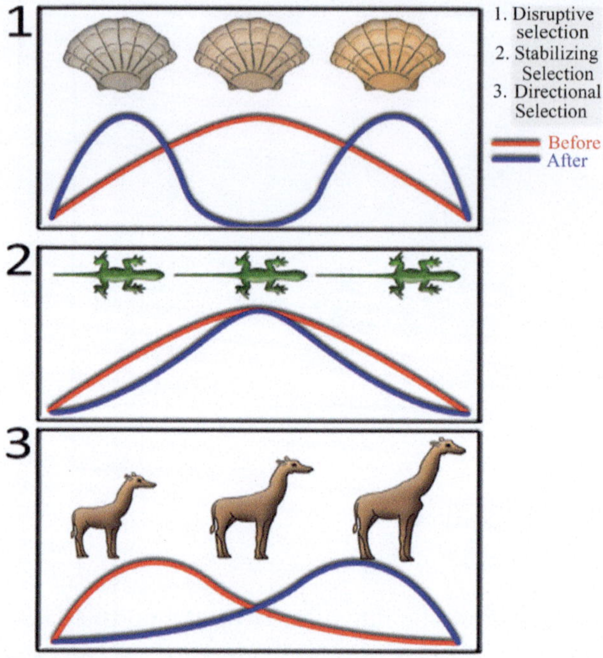

Figure 3.2. Different forms of natural selection.

Source: Image by commons.wikimedia.com.

Its gene frequencies and its total gene pool are two important attributes of a population recognized by the population growth. Since the number of sexually interbreeding individuals influences the transfer of genes within a population size exerts some influence on the genetic composition of its members.

The kinds of genotypes also get affected, that are the survival and available, and reproductive capacity of individuals with certain genes. To improve the breeds of animals and plants for agricultural use, the application of the principles of population genetics is vital to the success of programs and for the captive breeding of endangered species.

3.2.1. Definition of a Population

The total number of individuals of a given biological species found in one place at one time is called population. Ecologists often deal with weight rather than raw numbers or even with density (numbers per unit area for land organisms and numbers per unit volume in aquatic systems) in practice. Depends on the kind of organism and the aim of the scientific inquiry, what may be described as an "individual."

The life cycle starts with a fertilized egg, passes through a largely irreversible process of coupled growth and differentiation, and ends in a tightly integrated, unitary, adult organism in most animals. By merely counting adult units and their juvenile progeny population size can be measured.

> Growth and differentiation proceed in a modular fashion, in most plants and some colonial animals; growth involves the replication of a basic body unit, in these cases, a spatially distributed "population" of connected modules is generated by a fertilized module., In size and form modular organisms show tremendous plasticity, in general.

Birth, death, immigration, and emigration, there are only four ways that a population can change in size. The difference between birth and death rates drives its dynamics and if immigration and emigration are negligible, the population is closed. Terrestrial animals often have closed populations on islands.

However, the population is open but if immigration and emigration are important, and by spatially distant events its abundance may be substantially influenced. As for an example, on a rocky coastline the number of barnacles that are found often reflects the density of setting larvae, which in turn is governed in offshore waters by events.

To capture the important mechanisms of its population dynamics, if a population that is under study is found to be highly open, the spatial scale of the study may be too narrowly circumscribed.

3.2.2. Basic Population Models

Populations exhibit a great variety of dynamical patterns, including regular cycles or relatively constant abundances local extinctions, and explosive outbreaks. Ecologists rely on population models to help describe and explain these patterns. In developing population models' simple life cycles and closed populations provide a useful starting point.

For instance, one annual generation consisted by many temperate-zone insects have; thus, all individuals are at the same stage of life, at any given time. To project population numbers through time, iterations of discrete time-growth models for subsequent generations allow one. Including extensions to more complicated life cycles and multiple species, the theoretical framework of population ecology largely consists of elaborations of basic growth models.

Exponential growth has a snowballing effect in addition—as for an example, the faster that the population grows (if the growth rate is constant) if there are the more individuals. Enormous sizes population eventually lead by even low growth rates. Particularly in colonizing episodes, some natural populations show transient phases of exponential growth (Figure 3.3).

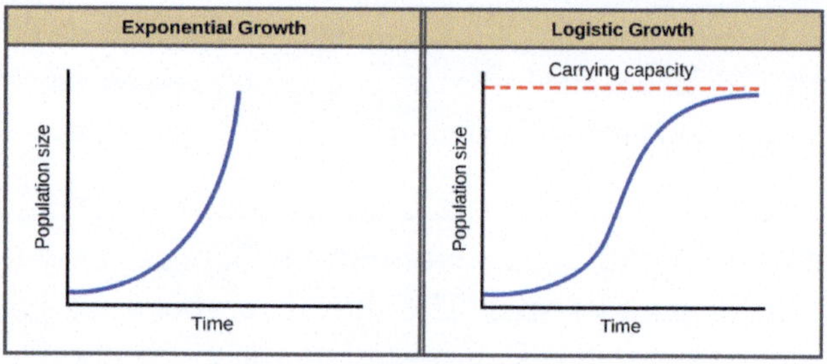

Figure 3.3. Exponential and logistic growth curves.

Source: Image by commons.wikimedia.com.

During colonization exponential growth often involves expansion across space, in addition and an increase through time. Thus, if appropriate, immigration, and emigration need to be considered.

3.2.3. Population Heterogeneity

When interpreted literally population models, that all members of a population are identical is assumed. This is rarely true, though. As a function of age, body weight, and genotype, birth, and death rates typically vary. Using these patterns to predict population growth, and interpreting these patterns in the light of evolutionary theory, a great deal of work in population ecology is devoted to elucidating age-specific schedules of mortality and fecundity.

The number of individuals in each age class must be monitored, to study age-structured population dynamics. To project changes in population size and age structure are the mortality schedule or survivorship curve, the two ingredients needed, which describes the rate of female births per female at each age, which describes the fraction of new-born's surviving to each age, and the fecundity schedule (Figure 3.4).

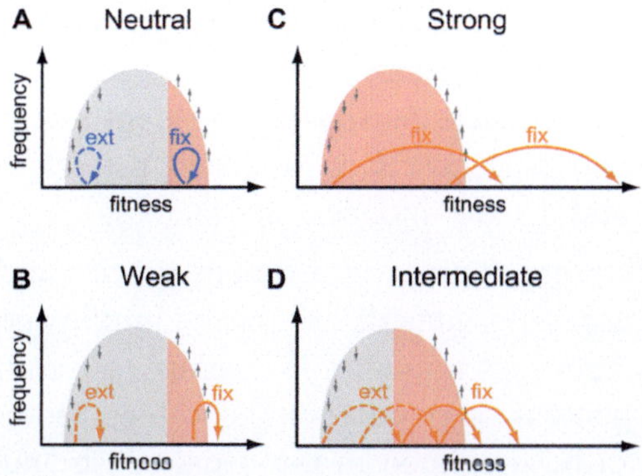

Figure 3.4. Fates of mutation in heterogeneous population.

Source: Image by commons.wikimedia.com.

In natural populations, it is a formidable task to measure complete fecundity and mortality schedules. Geometric growth models can be generalized to a matrix model, however, if these schedules are estimated. Moreover, a population will (with rare exceptions) eventually settle into a stable age distribution, if these schedules are constant, a constant fraction of the total population is comprised by each age.

From the mortality and fecundity schedules, a population in its stable age distribution grows geometrically at a rate of increase uniquely determined. Diverging from its long-term growth pattern, a population displaced from its stable age distribution may exhibit transient phases of growth or decline. All populations are genetically variable.

Genetic variation can influence population dynamics, if different genotypes have different fecundity or mortality schedules. Population growth can be quite complex and models that simultaneously incorporate changes in genetic composition. However, they are necessary for linking population ecology with evolutionary theory and may be important in describing some populations.

3.2.4. Population Limitation and Regulation

Populations cannot expand exponentially forever. As a result of competition for limited resources, often, mortality rates increase or birth rates decrease (or both). In these cases, the underlying mechanisms are called negatively density-dependent factors, which are specific examples of the more general concept of feedback.

Leading to population declines, alternatively, environmental conditions (for example, climatic shifts) can worsen. In birth or death rates, such causes for variation are called density-independent factors. In determining population size, ecologists have long disputed the relative importance of density-dependent and density-independent factors.

However, to differing degrees in different species and environments, it is likely that both are important. By no means implies population stability in population regulation, intriguingly. If it returns to equilibrium following a perturbation in general, a population is stable. Moreover, many local populations may not be regulated in the usual sense and thus may not be persistent over long periods of time.

As for an example, by definition coupled by dispersal with other populations, open populations, by immigration following a local extinction, that can become re-established. In determining population size is to plot birth and death rates as functions of density, a useful method for considering the interplay of density-dependent and density-independent factors.

The largest number of individuals for which the birth rate just matches the death rate is known as the carrying capacity of a population in a given environment. In general, density-independent factors must also be considered to understand fully what limits populations to a given value of carrying capacity but density-dependent factors are necessary to regulate populations.

3.2.4.1. Mechanisms of Density Dependence

The mechanisms generating, given that density dependence exists, it can be used both to provide insight into systems where experimental manipulations are difficult and to predict the consequences of environmental change for population dynamics. Density dependence often arises from competition, thereby negatively affect each other and which is said to exist when organisms utilize common limiting resources and therefore negatively affect each other.

Interference and exploitative, there are two principal sorts of competition. When one individual directly harms another interference competition occurs. As in lethal aggression, interference may be dramatic, or subtle, as when social interactions increase the risk of predation or reduce the time available for gathering resources.

Exploitative competition occurs when one individual consumes a resource, such as food, that otherwise would have been consumed by another individual. Because exploitative competition is mediated indirectly through a shared resource base, it can be more difficult to demonstrate than interference competition.

3.2.4.2. Population Regulation and Interspecific Interactions

From interspecific interactions, negative density dependence may arise. Between two species, a schematic classification of interactions comes from considering the positive (+) or negative (−) effect that individuals of one species have on the growth rate of the other. The interaction is (−, −); in mutualism, it is (+, +), in interspecific competition.

With their prey or hosts, natural enemies, defined broadly to include predators, herbivores, and parasites, are often engaged in (+, −) relations. Even top-level carnivores may be beset by parasites and most species are potential prey to one or more natural enemies. When studying population limitation, obviously, competitors, and mutualists can dramatically affect the size of a given population and thus must be considered.

However, than are either competitors or mutualists natural enemies are far more likely to be regulatory agents. One increases in density, the other will decrease, if two species are competing. Which can lead to increase even more, this will relax the interspecific competition on the first. Hence, competitive loops (and similarly mutualist loops) will not regulate population growth and tend to produce positive feedback and.

By contrast, negative density dependence acting across several time scales produced by the predator-prey interactions on both the predator and prey. Predators may almost immediately become more active or switch over from other prey types or patches, as prey in one habitat patch become more numerous.

An intergenerational numerical response to increased prey availability is shown by predators. Prey numbers usually decreased by the increase in predators, in both the predator and its prey, this induces delayed density dependence.

3.2.4.3. Applied Population Ecology

Population ecology has great practical utility, in addition to its intrinsic conceptual appeal. To reduce the intrinsic rate of increase of those organisms to very low values, control programs for agricultural pests or pathogens that cause human diseases ideally attempt. The development of vaccination programs is successfully guided by the analyzes of the population dynamics of infectious diseases.

Such as in forestry or fisheries biology in the exploitation of renewable resources, in order to devise sensible harvesting strategies, population models are required that maximize the sustainable yield extracted from exploited populations. With the consequences of habitat fragmentation for species preservation conservation biology is increasingly concerned.

Population models can help guide the development of interventionist policies to save endangered species and can help characterize minimum viable population sizes below which a species is vulnerable to rapid extinction. Finally, to bring the world's burgeoning human population into harmonious balance with the environment population ecology must be an integral part of any attempt.

3.2.4.4. Concepts and Principles of Population Dynamics

From other such groups, a population is a collection of individuals of a single species of organisms spatially or temporally isolated. Populations are geographically contiguous in some cases, but exhibit gradual genetic variation across their range (clinal variation).

In terms of the means and variances of parameters descriptive of its morphology, behavior, biological attributes, and distribution, in a qualitative sense, a population can be considered at various levels of spatial resolution and will exhibit different characteristics. Such parameters can be used to quantify degrees of difference or similarity, in comparing two or more populations within a species.

For example, Chitwood, is calculated from the mean for samples of the population taken from each state in which it occurs, the mean of all the populations will have a large associated variance if the thermal optimum for development of Meloidogyne incognita (Kofoid and White).

In a single state, if the mean is determined for the population, it will probably differ from the national average and have a much smaller variance. Similarly, if the mean is determined for the population in a single field, it will have a still smaller variance and vary from both state and national averages.

With the level of resolution at which they are measured, similarly, the spatial characteristics and distributional measures of a population vary. The perception of aggregation in the population increases as the physical size of the sample unit decreases. M. incognita may appear relatively uniformly distributed across the southern United States.

With regions of high population density, regions of low population density, and regions of apparent absence of the population within a single field, the nematode may appear to have an aggregated distribution. However, there are clearly islands of high aggregation (multiple females in single galls of the root producing egg masses) and large portions of the root and rhizosphere unoccupied by the nematode, at the level of resolution of a single plant root system.

The measurement of aggregation may be reversed towards randomness or further to uniformity as the physical resolution is defined at a still finer level (for example, distribution of eggs within a single egg mass). In that it parallels the information on population dynamics of a species obtained by experimental techniques varying in their resolution, this change in perception of aggregation with changing physical resolution is relevant.

In other words, a function of the magnitude of the system included in, or described by, these measures by many of the quantitative measures of populations. At the level of the population of a single species considered the boundary of the system for the purpose of this chapter.

Consequently, intrinsic properties of the population are considered by all factors within this boundary are considered. To the population and to the system, factors outside the boundary, including food resources, environmental conditions, and other biotic agencies are considered extrinsic.

To convey changes in the numbers, age class distribution, sex ratio, and behavior of a population through time and space the term population dynamics is used, mediated by environmental conditions and determined by inherent characteristics of the individuals, food resources, and interacting biotic agents. The basis of their dynamics and that are definable and measurable characteristics of the population. Age-specific rates of development, mortality, and reproduction exhibited by them. At a definable rate they migrate into or out of an area. They have a measurable sex and age composition which is also determinant of the dynamics of change in the population and which may or may not be stable.

3.2.4.5. Intrinsic Features of Populations

Nematodes, have definable life history strategies like other organisms. In terms of the number of life staged, these strategies can be defined the duration of the life stages, fecundity rates, life expectancy, sex ratios, and functionality of males. They can also be defined in the case of plant parasites, in terms of the feeding habits and mode of parasitism.

By density-dependent and density-independent factors, the strategies may be impacted. The survival of organisms, their reproductive potential and their longevity is impacted by the resource limitation. Temperature will affect consequently duration of life stages and fecundity rates and the rates of metabolic processes.

In terms of their life history strategy, we categorize organisms to aid in our understanding of the biology and adaptive significance of these strategies and in the development of concepts of the evolutionary. In response to the impact of both density-dependent and density-independent conditions, life history strategies may undergo tactical shifts.

Such alterations of the general strategies are factors that promote its survival and success in its environment and examples of the plasticity and adaptability of the organism. In terms of their feeding habits and mode of parasitism, life history strategies of plant-parasitic nematodes, involving varying levels of sophistication of the relationship between parasite and host, range from ectoparasite through various intermediate categories to sedentary endoparasite.

Involvement of an egg stage, four juvenile stages, and male and female adults, although males are frequently rare or may be absent, life history attributes of plant-parasitic nematodes. There is indication of a tendency to shortening or elimination of life stages, in many cases. Thus, some developmental stages may be relatively short in the first juvenile stage which is frequently passed in the egg.

In response to resource limitation or adverse density independent conditions include, tactical shift capabilities in life history strategies: sex ratio alteration, which may promote genetic recombination; parthenogenesis as an adaptation to the parasitic mode and which impacts the fecundity of the organism; diapause, aestivation, and cryptobiosis as mechanisms of survival of adverse conditions; width of host range and special mechanisms including host recognition.

Variability among individuals is important, at the population level. The basis for selection and adaptation to environmental change is provided by the variability at the genetic level. It mediates and allows the evolution of the organism. In terms of tactical variation in life history strategies and genetic breadth to provide germplasm adapted to changed conditions, plasticity, in response to environmental shifts can be measured.

3.2.4.6. Interaction with Extrinsic Regulators

Primary density dependent regulator of the population dynamics of nematodes and other organisms is a resource availability. In the soil matrix (horizontal and vertical), the abundance and state of the food source (its physiological activity and health), its distribution and its host status to the nematode all impact rates of population change.

Resource availability affects and interspecific competition and mediates intraspecific among individuals. In terms of temperature, soil moisture, aeration, pH, salinity or osmotic potential, particle or pore size affect and constrain developmental rates, activity, mortality, and fecundity, the status of the environment.

3.2.5. Population and Community Ecology

Imagine in a small motorboat on a weekend afternoon sailing down a river; you are enjoying the warm sunshine and the water is smooth and cool breeze when suddenly you are hit by a 20-pound silver carp in the head. This is a risk now on many rivers and canal systems because of the presence of Asian carp in Illinois and Missouri.

Actually, a group of species including the silver, black, grass, and big head carp, this fish—has been farmed and eaten for over 1000 years in China. Worldwide, it is one of the most important aquaculture food resources. However, Asian carp is considered a dangerous invasive species that disrupts community structure and composition to the point of threatening native species, in the United States.

3.3. POPULATION DEMOGRAPHY

Populations are dynamic entities. All of the species living within a specific area consisted by the population, and based on a number of factors populations fluctuate: yearly and seasonal changes in the environment, natural disasters such as volcanic eruptions, and forest fires and competition for resources between and within species.

To investigate how populations respond to changes in their biotic and abiotic environments, the statistical study of population dynamics, demography, uses a series of mathematical tools. To study human populations many of these tools were originally designed. For example, life insurance companies were initially developed life tables, which detail the life expectancy of individuals within a population, to set insurance rates. In fact, when discussing humans while the term "demographics" is commonly used, using this approach, all living populations can be studied.

3.3.1. Population Size and Density

By determining how many individuals of a particular species exist the study of any population usually begins, and how closely associated they are with each other. A population can be characterized by its population size (N), within a particular habitat, the total number of individuals, and its population density, the number of individuals within a specific volume or area.

To describe and understand populations, population size and density are the two main characteristics used. For instance, based on their genetic variability, populations with more individuals may be more stable than smaller populations, and thus to adapt to the environment their potential.

Alternatively, compared to a population of higher density, a member of a population with low population density (more spread out in the habitat), might have more difficulty finding a mate to reproduce.

3.3.2. Species Distribution

By looking at the distribution of the individuals, in addition to measuring simple density, further information about a population can be obtained. Within a habitat at a particular point in time, species dispersion patterns (or distribution patterns) between members of a population, show the spatial relationship. In other words, they show what patterns are evident when they are spaced apart and whether members of the species live close together or far apart.

In a population, individuals can be more or less equally spaced apart, in groups, dispersed randomly with no predictable pattern, or clustered. These are known as uniform, random, and clumped dispersion patterns. Inhibiting the growth of nearby individuals, uniform dispersion is observed in plants that secrete substances (by the sage plant *Salvia leucophylla*, such as the release of toxic chemicals a phenomenon called allelopathy) and in animals like the penguin that maintain a defined territory.

An example of random dispersion occurs with dandelion and other plants wherever they happen to fall in a favorable environment, that have wind-dispersed seeds that germinate. Such as oak trees, or animals that live in groups (schools of fish or herds of elephants), a clumped dispersion may be seen in plants that drop their seeds straight to the ground.

A function of habitat heterogeneity is clumped dispersions. Thus, how they interact with each other than does a simple density measurement is the information we get by the dispersion of the individuals within a population. Just as lower density species might have more difficulty finding a mate, when compared to social species clumped together in groups, solitary species with a random distribution might have a similar difficulty.

3.3.3. Demography

Scientists must use demography to study the dynamics of a population, while population size and density describe a population at one particular point in time. The statistical study of population changes over time is demography: birth rates, death rates, and life expectancies. By the population characteristics described above, each of these measures, especially birth rates, may be affected.

For example, because more potentially reproductive individuals are present a large population size results in a higher birth rate. In contrast, because of competition, disease, and the accumulation of waste, a large population size can also result in a higher death rate. Similarly, a clumped dispersion pattern or a higher population density result in more potential reproductive encounters between individuals, which can increase birth rate.

Lastly, age structure (at specific age ranges the proportion of population members) or a female-biased sex ratio (the ratio of males to females) or composed of many individuals of reproductive age can increase birth rates. In addition, how the population grows or declines over time influenced by the demographic characteristics of a population.

If the population remains stable, birth, and death rates are equal. However, if birth rates exceed death rates, the population size will increase; if birth rates are less than death rates, the population will decrease. Another important factor is life expectancy; local resources, reproduction, and the overall health of the population impacted by the length of time individuals remain in the population. In the form of a life table these demographic characteristics are often displayed.

3.4. LIFE HISTORIES AND NATURAL SELECTION

Such as how resources are allocated for growth, maintenance, and reproduction A species' life history describes the series of events over its lifetime. The life table of an organism is affected by life history traits. By the environment and natural selection, A species' life history is genetically determined and shaped.

3.4.1. Life History Patterns and Energy Budgets

Energy is often a major limiting factor in determining an organism's survival, energy is required by all living organisms for their growth, maintenance, and reproduction; at the same time. As for an example, from the sun via photosynthesis plants acquire energy, but to produce the next generation, plants must expend this energy to grow, maintain health, and produce energy-rich seeds.

To acquire food, animals have the additional burden of using some of their energy reserves. For their offspring, furthermore, some animals must expend energy caring. Thus, energy budget contains by all species: for metabolism, reproduction, parental care, and energy storage (such as bears building up body fat for winter hibernation) they must balance energy intake with their use of energy.

3.4.2. Parental Care and Fecundity

Within a population, fecundity is the potential reproductive capacity of an individual. In other words, fecundity describes if an individual has as many offspring as possible how many offspring could ideally be produced, as soon as possible after the birth of the offspring repeating the reproductive cycle.

The amount of parental care given to an individual offspring is inversely related to fecundity, in animals. Species, that produce many offspring usually provide little if any care for the offspring (they would not have the energy or the ability to do so anyway) such as many marine invertebrates. To produce many tiny offspring, most of their energy budget is used.

At a very early age, animals with this strategy are often self-sufficient. These organisms have made to maximize their evolutionary fitness this is because of the energy tradeoff. It makes sense that these offspring have some ability to be able to move within their environment and find food and perhaps shelter, because their energy is used for producing offspring instead of parental care.

The production of many offspring allows enough of them to survive to maintain the species because even with these abilities, their small size makes them extremely vulnerable to predation. Devoting much of their energy budget to these activities, sometimes at the expense of their own health animal species that have few offspring during a reproductive event usually give extensive parental care.

Such as humans, kangaroos, and pandas, this is the case with many mammals. Before they achieve self-sufficiency, the offspring of these species are relatively helpless at birth and need to develop. With each having a good chance to germinate into a new organism; plants with high fecundity usually have many small, energy-poor seeds (like orchids) that have a relatively poor chance of surviving, plants with low fecundity produce few energy-rich seeds (such as coconuts and chestnuts).

The energy tradeoff of the orchid is also very effective, although it may seem that coconuts and chestnuts have a better chance of surviving. For large numbers of seeds or for fewer seeds with more energy, it is a matter of where the energy is used.

3.4.3. Early Versus Late Reproduction

In a life history, the timing of reproduction also affects species survival. At an early age have a greater chance of producing offspring organisms that reproduce, but this is usually at the maintenance of their health and the expense of their growth.

Conversely, organisms that risk that they will not survive to reproductive age but they start reproducing later in life often have greater fecundity or are better able to provide parental care. Examples of this can be seen in fishes.

To reproduce rapidly, small fish like guppies use their energy, but never attain the size that would give them defense against some predators. Larger fish, use their energy to attain a large size, like the bluegill or shark, but they will die before they can reproduce or at least reproduce to their maximum this is the risk involved. To understanding the evolution of each species these different energy strategies and tradeoffs are key, as it maximizes its fitness and fills its niche. Some species "blow it all" and use up most of their energy reserves to reproduce early before they die, in terms of energy budgeting. To make sure that they are strong enough to provide parental care if necessary other species delay having reproduction to become stronger, more experienced individuals.

3.4.4. Environmental Limits to Population Growth

Population ecologists make use of a variety of methods to model population dynamics mathematically, although life histories describe the way many characteristics of a population (such as their age structure) change over time in a general way.

To accurately describe changes occurring in a population and better predict future changes these more precise models can then be used. Due to their lack of predictive ability, and scholars strive to create effective new models, certain models that have been accepted for decades are now being modified or even abandoned.

3.4.5. Exponential Growth

By the English clergyman Thomas Malthus influences greatly Charles Darwin, in his theory of natural selection. A book was published by Malthus in 1798 stating that population growth decreases as resources become depleted and then populations with unlimited natural resources grow very rapidly.

Exponential growth is accelerating pattern of increasing population size. The best example of exponential growth is seen in bacteria. By prokaryotic fission, bacteria are prokaryotes that reproduce. For many bacterial species, this division takes about an hour.

There is one round of division after an hour and each organism divides, resulting in 2000 organisms—an increase of 1000, if 1000 bacteria are placed in a large flask with an unlimited supply of nutrients (so the nutrients will not become depleted).

In another hour, an increase of 2000 organisms, each of the 2000 organisms will double, producing 4000. After the third hour, an increase of 4000 organisms, there should be 8000 bacteria in the flask. The population growth rate is the important concept of exponential growth—in each reproductive generation the number of organisms added—is accelerating; that is, at a greater and greater rate it is increasing.

The population would have increased from 1000 to more than 16 billion, after 1 day and 24 of these cycles. A J-shaped growth curve is produced, when the population size, N, is plotted over time.

3.4.6. Logistic Growth

This is not the case in the real-world exponential growth is possible only when infinite natural resources are available. In his description of the "struggle for existence," Charles Darwin recognized this fact which states that individuals will compete for limited resources (with members of their own or other species). To pass on their own characteristics and traits, the successful ones will survive (which we know now are transferred by genes) to the next generation (natural selection) at a greater rate. Population ecologists developed the logistic growth model to model the reality of limited resources.

3.5. POPULATION DYNAMICS AND REGULATION

Simplification of real-world population dynamics is the logistic model of population growth, while valid in many natural populations and a useful model. The carrying capacity of the environment does not change in implicit in the model, which is not the case.

The carrying capacity varies annually: for example, some summers are cold and wet whereas others are hot and dry. In many areas, during the winter the carrying capacity is much lower than it is during the summer. Also, such as volcanoes, earthquakes, and fires, natural events can alter an environment and hence it carrying capacity.

Additionally, populations do not usually exist in isolation. In interspecific competition they are engaged: that is, they share the environment with other species, for the same resources competing with them. To understanding how a specific population will grow these factors are also important. In a variety of ways nature regulates population growth.

These are grouped into density-dependent factors, in which at a given time affects growth rate and mortality the density of the population, and density-independent factors, which influence mortality in a population regardless of population density.

Note that in the former, on the density of the population at onset the effect of the factor on the population depends. Because this helps them manage populations and prevent extinction or overpopulation conservation biologists want to understand both types.

3.5.1. Density-Dependent Regulation

In nature (biotic) most density-dependent factors are biological, and include accumulation of waste predation, inter-, and intraspecific competition, and diseases such as those caused by parasites. Usually, the denser a population is, the greater its mortality rate. For example, the reproductive rates of the individuals will usually be lower, during intra-and interspecific competition, reducing their population's rate of growth. In addition, because it has more difficulty locating its food source, low prey density increases the mortality of its predator.

From a study focusing on the giant intestinal roundworm (Ascaris lumbricoides) an example of density-dependent regulation, a parasite of humans and other mammals. Lower fecundity exhibits: they contained fewer eggs denser populations of the parasite.

Females would be smaller in more dense populations (due to limited resources) is the one possible explanation for this is that and that smaller females would have fewer eggs. In a 2009, this hypothesis was tested and disproved, the study which showed that female weight had no influence. In this organism is still unclear and awaiting further investigation the actual cause of the density-dependence of fecundity.

3.6. DENSITY-INDEPENDENT REGULATION AND INTERACTION WITH DENSITY-DEPENDENT FACTORS

Typically, physical or chemical in nature (abiotic) many factors, influence the mortality of a population regardless of its pollution, density, including weather, and natural disasters. Regardless of how many deer happen to be in that area, an individual deer may be killed in a forest fire.

Whether the population density is high or low its chances of survival are the same. The same holds true for cold winter weather. In real-life situations, density-dependent, and independent factors can interact and population regulation is very complicated.

By some environmental factor (s), a dense population that is reduced in a density-independent manner will be able to recover differently than a sparse population. For example, if there are more deer remaining to reproduce a population of deer affected by a harsh winter will recover faster.

3.6.1. Human Population Growth

Human population growth applies to the concepts of animal population dynamics. To alter their environment humans are not unique in their ability. For example, beaver dams alter the stream environment where they are built.

To increase it carrying capacity sometimes to the detriment of other species, humans, however, have the ability to alter their environment (e.g., via artificial selection for crops that have a higher yield). To the extent that some worry about the ability of the earth's environment to sustain this population, earth's human population is growing rapidly, as long-term exponential growth carries the potential risks of large-scale death, famine, and disease.

The technologies used to achieve this transformation have caused unprecedented changes to Earth's environment, altering ecosystems to the point where some may be in danger of collapse, although humans have increased the carrying capacity of their environment.

Human activities cause the depletion of the ozone layer, damage from global climate change and erosion due to acid rain. On our carrying capacity, the ultimate effect of these changes is unknown. It is likely that the negative effects of increasing carrying capacity will outweigh the positive ones, as some point out—the carrying capacity of the world might actually decrease for human beings.

Even though human reproduction is far below its biotic potential the world's human population is currently experiencing exponential growth. All females would have to become pregnant every nine months or so during their reproductive years to reach its biotic potential.

The environment would support such growth, also, resources would have to be such. These two conditions do not exist. Human population is still growing exponentially, in spite of this fact.

3.6.2. Overcoming Density-Dependent Regulation

With the conscious purpose of increasing, it carrying capacity, humans are unique in their ability to alter their environment. For human population growth and a way of overcoming density-dependent growth regulation this ability is a major factor responsible.

To human communication, intelligence, and society, much of this ability is related. To protect them from the elements, humans can construct shelter and to increase their food supplies, have developed agriculture and domesticated animals. In addition, to communicate this technology to new generations, humans use language, allowing them to improve upon previous accomplishments.

In human population growth other factors are public health and migration. In Africa humans originated, but to nearly all inhabitable land on the Earth, has since migrated. The ability of infectious disease to limit human population growth, public health, sanitation, and the use of antibiotics and vaccines have decreased. In the past, diseases of the fourteenth century such as the bubonic plaque reduced the overall world population by as many as 100 million people and killed between 30 and 60% of Europe's population.

Today, the threat of infectious disease, is certainly less severe while not gone. Global death from infectious disease declined from 16.4 million in 1993 to 14.7 million in 1992, according to the World Health Organization.

The percentage of the world's population killed between 1993 and 2002 decreased from 0.30% of the world's population to 0.24% to compare to some of the epidemics of the past. Thus, the influence of infectious disease on human population growth is becoming less significant, it appears that.

3.6.3. Age Structure, Population Growth, and Economic Development

In population dynamics the age structure of a population is an important factor. At different age ranges age structure is the proportion of a population. Better prediction of population growth allowed by age structure, plus with the level of economic development in the region, the ability to associate this growth. In their age structure diagrams countries with rapid growth have a pyramidal shape, many of whom are of reproductive age or will be soon, showing a preponderance of younger individuals. In underdeveloped countries this pattern is most often observed because of less-than-optimal living conditions where individuals do not live to old age.

Including developed countries age structures of areas with slow growth, such as the United States, still have a pyramidal structure, but a greater proportion of older individuals and with many fewer young and reproductive-aged individuals.

Such as Italy, have zero population growth like other developed countries. With an even greater percentage of middle-aged and older individuals the age structure of these populations is more conical. In different countries, the actual growth rates with the highest rates tending to be in the less economically developed countries of Africa and Asia.

3.6.4. Long-Term Consequences of Exponential Human Population Growth

"Population explosion" many dire predictions have been made about the world's population leading to a major crisis. Biologist Dr. Paul R. Ehrlich wrote, "The battle to feed all of humanity is over. In the 1970s hundreds of millions of people will starve to death in spite of any crash programs embarked upon now.

At this late date nothing can prevent a substantial increase in the world death rate." in the 1968 book The Population Bomb. As an exaggeration while many critics view this statement, unchecked human population growth cannot continue indefinitely and the laws of exponential population growth are still in effect.

Which used to include more severe consequences, efforts to control population growth led to the one-child policy in China, but now imposes fines who have more than one child on urban couples. Many Chinese couples continue to have more than one child, due to the fact that some couples wish to have a male heir.

The policy, its social impacts, and the overall population growth are controversial. The human population continues to grow, in spite of population control policies. Because of the subsequent need to produce more and more food to feed our population at some point the food supply may run out.

By the year 2100, the future world population growth may vary from 6 billion (a decrease) to 16 billion people estimated by the United Nations. Described by Dr. Ehrlich will be averted that there is no way to know whether human population growth will moderate to the point where the crisis.

Endangerment of the natural environment is another result of population growth. To reduce the human impact on climate change by reducing their emission of the greenhouse gas carbon dioxide many countries have attempted.

However, by every country these treaties have not been ratified, and if it means slower economic development, many underdeveloped countries trying to improve their economic condition may be less likely to agree with such provisions.

Furthermore, in some developed countries, including the United States, the role of human activity in causing climate change has become a hotly debated socio-political issue. Thus, to curb human population growth and protect our environment, we enter the future with considerable uncertainty about our ability.

3.7. COMMUNITY ECOLOGY

Live in isolation from populations of other species populations rarely, if ever. In most cases, numerous species share a habitat. A major role in regulating population growth and abundance played by the interactions between these populations.

The same habitat forms a community occupied by all the populations: populations inhabiting a specific area at the same time. Species diversity is the number of species occupying the same habitat and their relative abundance.

Such as the glaciers of Antarctica, still contain a wide variety of living things, areas with low diversity, whereas the diversity of tropical rainforests is so great that it cannot be counted. To understand how species interact with each other and compete for the same resource's ecology is studied at the community level.

3.7.1. Predation and Herbivory

Predation is perhaps the classical example of species interaction: by its predator the hunting of prey. The drama of one living organism killing another nature shows on television highlight. Populations of predators and prey in a community are not constant over time: they vary in cycles that appear to be related, in most cases.

Using nearly 200-year-old trapping data from North American forests, the most often cited example of predator-prey dynamics is seen in the cycling of the lynx (predator) and the snowshoe hare (prey). With the predator population lagging 1–2 years behind that of the prey population this cycle of predator and prey lasts approximately 10 years.

There is more food available for the lynx, as the hare numbers increase, allowing the lynx population to increase as well. However, the lynx population kill so many hares that hare population begins to decline, when the lynx population grows to a threshold level, because of scarcity of food followed by a decline in the lynx population. The hare population size begins to increase due, at least in part, to low predation pressure, when the lynx population is low, starting the cycle anew.

By predation models has come under question the idea that the population cycling of the two species is entirely controlled. In addition to predation, as being important in the cycling, more recent studies have pointed to undefined density-dependent factors.

Due to density-dependent effects, one possibility is that the cycling is inherent in the hare population such as lower fecundity (maternal stress) caused by crowding when the hare population gets too dense. Because it is the lynxes' major food source the hare cycling would then induce the cycling of the lynx.

Allowing ecologists to derive more accurate and sophisticated models of population dynamics, the more we study communities, the more complexities we find. By insects and other animals, herbivory describes the consumption of plants and that affects populations it is another interspecific relationship.

Unlike animals, most plants cannot outrun predators or use mimicry to hide from hungry animals. To defend against herbivory some plants have developed mechanisms. Other species have developed mutualistic relationships; for example, a mechanism of seed distribution that aids in plant reproduction provided by herbivory.

3.7.2. Defense Mechanisms against Predation and Herbivory

The study of communities must consider evolutionary forces that act on the members of the various populations contained within it. Species are not static, but by natural selection and other evolutionary forces, slowly changing and adapting to their environment.

To escape predation and herbivory species have evolved numerous mechanisms. These defenses may be mechanical, physical, behavioral or chemical. Mechanical defenses, such as the presence of thorns on plants or the hard shell on turtles, herbivory, and discourage animal predation by causing physically preventing the predator from being able to eat the prey or by physical pain to the predator. Such as the foxglove which is extremely toxic when eaten, chemical defenses are produced by many animals as well as plants.

3.7.3. Competitive Exclusion Principle

Multiple species may compete to obtain them and resources are often limited within a habitat. In the ecosystem, all species have an ecological niche, which describes how they interact with other species in the community and how they acquire the resources they need.

Two species cannot occupy the same niche in a habitat stated by the competitive exclusion principle states. In other words, if they are competing for all the same resources different species cannot coexist in a community.

They both thrive, when grown individually in the laboratory. Leading to the latter's eventual extinction but when they are placed together in the same test tube (habitat), P. aurelia outcompetes P. causatum for food. If a population evolves this exclusion may be avoided to make use of a different resource, feeds during a different time of day, a different area of the habitat, or called resource partitioning. Different microinches are said to be occupied by the two organisms. By minimizing direct competition these organisms coexist.

3.7.4. Community Dynamics

Community dynamics are the changes in community structure and composition over time. Sometimes these changes are induced by environmental disturbances such as volcanoes, earthquakes, storms, fires, and climate change. Communities with a stable structure are said to be at equilibrium.

Following a disturbance, the community may or may not return to the equilibrium state. Succession describes the sequential appearance and disappearance of species in a community over time. In primary succession, newly exposed or newly formed land is colonized by living things; in secondary succession, part of an ecosystem is disturbed and remnants of the previous community remain.

3.7.5. Primary Succession and Pioneer Species

When new land is formed or rock is exposed, primary succession occurs: for example, such as those on the Big Island of Hawaii following the eruption of volcanoes. New land is continually being formed as lava flows into the ocean.

Approximately 32 acres of land is added each year on the Big Island. First, For the establishment of certain hearty plants and lichens with few soil requirements, weathering, and other natural forces break down the substrate enough known as pioneer species.

To further break down the mineral rich lava into soil where other these species help, less hardy species will grow and eventually replace the pioneer species. As these early species grow and die in addition, they add to an ever-growing layer of decomposing organic material and contribute to soil formation. With a set of organisms quite different from the pioneer species, over time the area will reach an equilibrium state.

3.7.6. Secondary Succession

In oak and hickory forests cleared by wildfire, a classic example of secondary succession occurred. Wildfires kill those animals unable to flee the area and will burn most vegetation. In the form of ash, their nutrients, however, are returned to the ground.

Thus, Due to severe fires, even when areas are devoid of life, to take hold the area will soon be ready for new life. The vegetation was dominated by tall trees before the fire, with access to the major plant energy resource: sunlight. While also shading the ground and other low-lying species their height gave them access to sunlight.

Though, these trees are no longer dominant after the fire. Thus, by quickly growing and spreading grasses and other pioneer species the first plants to grow back are usually annual plants followed within a few years. Due to, at least in part, other species, over many years, changes in the environment brought on by the growth of the grasses and shrubs will emerge along with small pine, oak, and hickory trees.

These organisms are called intermediate species. The forest will reach its equilibrium point where species composition is no longer changing and resembles the community before the fire eventually, over 150 years. As the climax community this equilibrium state is referred to, until the next disturbance which will remain stable.

3.8. INNATE BEHAVIORS: MOVEMENT AND MIGRATION

On response to stimuli innate or instinctual behaviors rely. A reflex action, the simplest example of this is, to stimulus an involuntary and rapid response. A doctor taps the patellar tendon below the kneecap with a rubber hammer to test the "knee-jerk" reflex.

The reflex of extending the leg at the knee the stimulation of the nerves there leads. This is similar to the reaction of someone who touches a hot stove and instinctually pulls his or her hand away. Even humans, still exhibit a variety of innate behaviors, with our great capacity to learn.

3.8.1. Innate Behaviors: Living in Groups

Even those that live relatively solitary lives, with the exception of those that can reproduce asexually, must mate because not all animals live in groups.

One animal signaling another so as to communicate the desire to mate usually involves in mating.

Mating rituals occurs when there are several types of energy-intensive behaviors or displays associated with mating. In terms of which animal benefits from the behavior other behaviors found in populations that live in groups are described.

In selfish behavior, only the animal in question benefits; one animal's actions benefit another animal; in altruistic behavior, cooperative behavior describes when both animals benefit. Some sort of communication between population members involves all of these behaviors involve.

3.8.2. Simple Learned Behaviors

The majority of the behaviors previously discussed were innate or at least have an innate component (on the innate behaviors variations may be learned). They are inherited and the behaviors do not change in response to signals from the environment.

Even though they may have instinctive components, conversely, learned behaviors, allow an organism to adapt to changes in the environment and by previous experiences it is modified. Habituation and imprinting are simple learned behaviors included, and to the maturation process of young animals it is important.

3.8.3. Conditioned Behavior

A stimulus becomes associated with a consequence where conditioned behaviors are types of associative learning. By its consequences, the behavioral response is modified with regards to its form, strength, or frequency, during operant conditioning.

3.8.4. Cognitive Learning

For humans and other intelligent animals, classical, and operant conditioning are inefficient ways to learn. By imitating the behavior of others and by taking instructions, some primates, including humans, are able to learn. By humans has made cognitive learning the development of complex language, the most prominent method of human learning, the manipulation of information using the mind. In fact, by reading this book that is how students are learning right now. As students read, they can make mental images of objects or organisms and imagine changes to them, or anticipate the consequences, behaviors by them.

By remembering past experiences, in addition to visual processing, cognitive learning is also enhanced, hearing sounds, tasting food, a variety of other sensory-based inputs and touching physical objects. Cognitive learning can be used to understand conditioning in detail because it is so powerful. Conditioning cannot help someone learn about cognition, in the reverse scenario.

By Wolfgang Köhler with chimpanzees, classic work on cognitive learning was done. By showing that they could learn how to solve a puzzle, he demonstrated that these animals were capable of abstract thought. In their cage when a banana was hung too high for them to reach, and randomly on the floor several boxes were placed, some of the chimps were able to stack the boxes one on top of the other, climb on top of them, and reach the banana. They could visualize the result of stacking the boxes even before they had performed the action it is implied by this. This type of learning is much more versatile and powerful than conditioning.

Although cognitive learning is the most efficient in using it, is not limited to primates. Maze running experiments done with rats were the first to show cognitive skills in a simple manner by H. C. Blodgett in the 1920s.

To work their way through the maze was a piece of food at its end was the motivation for the animals. In these studies, in Group I the animals were run in one trial per day and on completion of the run, had food available to them each day.

In Group II for the first six days rats were not fed in the maze and for several days after then subsequent runs were done with food. In Group III rats had food available on the third day and every day thereafter. The results were that the control rats, In Group I, rats learned quickly in seven days, and figured out how to run the maze.

In Group III rats did not learn much without food during the three days, but when given the food reward rapidly caught up to the control group. Group II learned very slowly with no reward to motivate them for the six days, and they did not begin to catch up to the control group until the day food was given, and to learn the maze it took them two days longer.

This type of learning is different than conditioning it may not be immediately obvious. Although one might be tempted to believe that the rats simply learned how to find their way through a conditioned series of right and left turns, E. C. Tolman proved a decade later that the rats were making a representation of the maze in their minds, which he called a "cognitive map." This was an early demonstration of the power of cognitive learning and how these abilities were not just limited to humans.

3.8.5. Sociobiology

By social insect researcher E. O. Wilson in the 1970s sociobiology is an interdisciplinary science originally popularized. "The extension of population biology and evolutionary theory to social organization" Wilson defined the science.

In terms of genetics and natural selection the main thrust of sociobiology is that animal and human behavior, including aggressiveness and other social interactions, can be explained almost solely. Noted scientist such as the late Stephen Jay Gould criticized the approach for ignoring the environmental effects on behavior, as this science is controversial. Debate of the role of genetics versus the role of environment, this is another example of the "nature versus nurture."

The belief that all behaviors are hardwired into our genes, sociobiology also links genes with behaviors and has been associated with "biological determinism,." That natural selection plays a role retaining them no one disputes those certain behaviors can be inherited. It is the application of such principles to human behavior that sparks this controversy, which remains active today.

3.9. CONCLUSION

The competition between species that fight for scarce resources attributes the characteristic change in population of organisms can significantly. This competition shapes causes extinction of some species and the biological population.

Including the possibility of such competition leading to extinction, the objective of this paper is to understand how competition between different species is important in the shaping of biological populations. Broadly, the paper deals with the survival of fittest, the extinction of the weak ones and the competition process between different species.

When resources are common and scarce, competition between species becomes inevitable. In real life it is a tendency to become scarce when population rises and resources are always limited. A limited resource that supports the growth of most of the species are carrying capacity of the earth, water, sunlight, and food are the common examples.

When these resources are in abundance and other growth conditions are favorable population growth is rapid. From less favorable environment leading to increase in population in fact, favorable growth conditions cause immigration of species. Emigration of species caused by discouraging external. Fewer resources are used up when the population is sparse.

REFERENCES

1. (n.d). *Population and Community Ecology*. [e-Book] http://rms.rsccd.edu/. Available at: http://rms.rsccd.edu/faculty/kimomorris/bio212/oer/oerbio_ch45.pdf (accessed on 27 June 2022).
2. Cumberlidge, N., (n.d). *Population Biology* | Encyclopedia.com. [online] Encyclopedia.com. Available at: https://www.encyclopedia.com/environment/encyclopedias-almanacs-transcripts-and-maps/population-biology (accessed on 27 June 2022).
3. Ferris, H., & Wilson, L., (n.d). *Concepts and Principles of Population Dynamics*. [e-Book] https://beaumont.tamu.edu/. Available at: https://beaumont.tamu.edu/eLibrary/Publications/Ted_Wilson/LTW73.pdf (accessed on 27 June 2022).
4. Ma, F., Liu, G., & Zhang, K., (2021). Goal-conditioned behavioral cloning with prioritized sampling. In: *2021 IEEE International Conference on Networking, Sensing and Control (ICNSC)*. [online] Available at: https://www.researchgate.net/search/publication?q=Conditioned%20Behavior%20 (accessed on 27 June 2022).

CHAPTER 4

Social Mechanisms

CONTENTS

During the last two decades, the social sciences and philosophy of science have paid close attention to social mechanisms and mechanism-based explanations. The concept of mechanical explanation has proven to be an effective instrument for evaluating present research techniques and metatheoretical perspectives on the nature of the social-scientific enterprise. Many different definitions of social processes have been proposed and utilized to support a wide range of methodological and theoretical assertions. Because it is difficult to address all of these in a single chapter, only the most prominent and philosophically significant topics were included.

4.1. INTRODUCTION

The idea that one variable is a cause of the other or that both are consequences of a shared cause is a basic challenge in social science. Given the general inability of conducting trials and the difficulty of knowing whether all probable common factors have been considered, establishing valid judgments regarding the origins of social phenomena is severely hampered.

It is frequently asserted that taking into account social factors might greatly alleviate this difficulty. In this paper, it was investigated whether mechanisms can execute this function and, if so, how the case for mechanisms is frequently made in terms of the notion that valid causal inference in social science is impossible without knowledge of mechanisms, which Harold Kincaid has challenged.

Although agreed with Kincaid that processes are not necessarily required for causal inference in social science, also believe that the suggestion may be made irrespective of that notion. The perspective, the description of how mechanisms to aid causal inference in social science contains both good and negative aspects.

On the plus side, we can conclude that X is a cause of Y if we know that X affects Y via a mechanism. The disadvantage is that if no plausible mechanism connecting X and Y can be imagined, it is reasonable to assume that X does not affect Y, even though the two variables are probabilistically linked. As demonstrated, neither of these lines of reasoning implies that mechanisms are required for causal inference in social science, and hence they are not undercut by Kincaid's arguments.

Nonetheless, it was contended that as it is, this theory of how processes significantly aid causal inference in social science is ineffective. The positive account is ineffective unless some explanation is provided as to how it is feasible to learn about social mechanisms while avoiding the difficulty of unmeasured common causes. To the best of my knowledge, no such explanation has been supplied thus far.

On the other hand, the simplicity with which processes connecting practically any two variables reflecting features of social phenomena may be imagined undermines the effectiveness of the negative side. However, it was believed that the account's positive side may be strengthened, resulting in a better explanation of the importance of mechanisms to causal inference in social science.

Author suggests that knowledge of social norms and behaviors may be gained independently of the problem of unmeasured common causes Moreover, by investigating how many social behaviors are connected together, one may sometimes construct a compelling case for a qualitative causal hypothesis stating that one variable is a cause of another. Several examples from social science, particularly anthropology, are used to support this thesis.

4.1.1. What Are Social Mechanisms?

A few of the first explicit applications of the word mechanism in the context of social science were in Robert Merton's 1949 article "On Sociological Theories of the Middle Range." Merton argued in that article for a social theory that avoids big systems that aim to cover all elements of human civilization. Instead, he advocated that social theory concentrate on theories about the dynamics of certain types of social interactions.

According to Merton, middle-range theories rely primarily on mechanisms. One of Merton's most well-known processes is the self-fulfilling prophecy, which he demonstrated using the example of a bank run and, more interestingly, the self-reinforcing components of racial prejudice.

Despite Merton's popularity, the notion that mechanisms, the concept of mechanisms that should have been important to social theory did not catch on in sociology, and little effort was made to define the concept. As a result, the philosophical literature on social processes is limited, owing mostly to a small number of philosophers.

However, there has recently been a resurgence of interest in the issue, and this revived interest in social processes has had an impact on the real study. One of the main reasons for interest in social mechanisms is the belief that adequate explanations in general, and of social phenomena in particular, necessitate an understanding of underlying causal processes in addition to the recognition of patterns of correlation and dependence among aggregate-level variables.

This portion of the literature on social mechanisms, however, is not the topic of this paper, which investigates the claim that social mechanisms provide an effective framework for alleviating the problem of confounders. But, before we get started, it's important to define the term "social mechanism."

Mechanisms in general may be defined as groups of things and activities that are structured in such a way that they create a regular series of changes from a starting state to an ending one. Social processes, in particular, are commonly seen as complexes of interpersonal interactions that underpin and account for aggregate social regularities.

However, social systems are more than just individual interactions: Usually, Individuals are classified into relevantly comparable groups based on a prominent position their members have in relation to other members of society. The relevant behavior of an individual is frequently considered to be a function of the group into which he or she is categorized in the mechanism's description.

Consider anthropologist Bronislaw Malinowski's story of how having more wives led to greater riches among Trobriand leaders. 4 Men among the Trobrianders were obligated to pay significant yearly yam donations to the households of their married sisters. As a result, the more wives a guy had, the more yams he received. Meanwhile, yams were the principal source of money in Trobriand culture, and they were used to fund activities such as canoe building and warfare.

Individuals play an important part in this story, but they do so as representations of societal categories: brothers-in-law, spouses, and chiefs. Social processes are not the only ones that categories component items into functionally defined kinds. Biological systems (for example, HIV replication) are frequently explained using terminology like enzyme and coreceptor.

Because they are functional, the phrases enzyme and coreceptor are similar to chief and brother-in-law: all of these terms convey some information about what role the designated entity performs in the wider system of which it is a part. The presence of social categories into which the persons forming the mechanism are classified is significant since social positions typically last considerably longer than the individuals who hold them, as demonstrated by the Trobriand example.

Stable social roles are therefore a source of consistency for the processes in which they are engaged, as well as for the higher-level causal generalizations that rely on these mechanisms. Social mechanisms are sometimes associated with the premise that the persons that comprise them are rational, as in utility maximizers. Tyler Cowen, for example, states, "I understand social processes as rational-choice explanations for how a given set of preferences and constraints might result in more complicated social consequences."

4.2. MECHANISMS AND THE PROBLEM OF CONFOUNDERS

A large portion of social science is concerned with gathering aggregate data relevant to some phenomenon of interest (e.g., through government records or surveys) and performing statistical tests to determine whether pairs of variables are probabilistically dependent conditional on sets of other variables.

In some circumstances, the objective of such investigations is simply to find elements that might be used as forecasting tools, but more commonly, the goal is to determine which variables affect which others. In the social sciences, this immediately leads to the difficult challenge of generating causal conclusions without the use of experimentation.

The main impediment to such findings is that a probabilistic dependence between two variables can be explained by either one variable being a cause of the other or both having the same source. This is known as the confounders issue because confounders are common, unmeasured causes that may explain an apparent association.

4.2.1. Kincaid's Objections

Even though the suggestion given in the preceding section claimed that investigations into social mechanisms can considerably alleviate the problem of confounders, no claim was made that they are required to resolve this problem. That assertion, however, is occasionally made in conjunction with the proposition outlined above.

According to Little, "social scientists will be able to employ empirical evidence to establish causal relationships solely on the basis of theories about underlying causal mechanisms." Kincaid makes two arguments against the idea that social processes are required for causal inference in the social sciences, the first of which is a reductio ad absurdum.

The notion at stake in Kincaid's (debate is that "we need to determined individualist processes are used to validate causal relationships between social factors."

4.2.2. No Plausible Mechanism

As previously stated, there are both good and negative aspects to the notion of the centrality of mechanisms to causal inference in the social sciences. The positive side is based on the assumption that we can demonstrate that X is a cause of Y if we can uncover a mechanism from X to Y.

The disadvantage is based on the assumption that we may deduce that X is not a cause of Y if we know there is no credible route from X to Y. Little's example of nicotine stains and lung cancer, Hedström and Swedberg's example of electromagnetic fields and infant leukemia, and Elster's example of the proportion of female employees and number of lost workdays all demonstrated the negative side.

In this part, I show that the negative side of the explanation of the role of mechanisms in causal inference in social science fails. The difficulty arises from the ease with which practically any aggregate-level social variable may impact another. As a result, it is unusual that no conceivable mechanism that could connect two variables reflecting features of social processes can be envisaged.

Consider, for example, a well-known case from the sociological literature mentioned by one of the volume's contributors, Diego Gambetta. The association is an example, as recorded in Samuel Stouffer's. The American Soldier a conflict between job satisfaction and promotion opportunities for military members.

Surprisingly, troops in parts of the military with little opportunities, such as the military police, were more content with their jobs than those in areas with more opportunities, such as the army air corps. Gambetta discusses five factors hypothesized by sociologists over the years to explain how increased opportunity might lead to decreased contentment. He does not, however, address the possibility that opportunity has little or no negative impact on happiness and that the link shown by Stouffer is due to an unmeasured shared factor. It is probable, for example, that exceptionally ambitious people are far more inclined to choose job routes that promise higher prospects for growth, and that their lofty ambitions are also more likely to leave them unsatisfied with their existing stations in life.

Listing different methods through which opportunity might lead to sadness does not rule out this conceivable possibility. Indeed, this scenario demonstrates how an oversupply of possible processes is a significant source of difficulties in causal inference in the social sciences.

There exist, beyond a doubt, some pairings of variables X and Y that describe aggregate features of social processes such that no logical mechanism by which X produces Y can be envisioned nevertheless, I feel that such examples are too far and few between for the no-plausible mechanism technique to be useful in social research in differentiating causation from the simple association.

Although Elster, Little, Hedström, and Swedberg all provide examples to support their points, only Elster's (a fictitious example not based on actual research) is relevant to social science. Despite their enthusiasm, these writers appear to have found it difficult to provide a genuine example of actual social research in which the difficulty to construct a credible process from one social variable to another considerably improved causal inference (Figure 4.1).

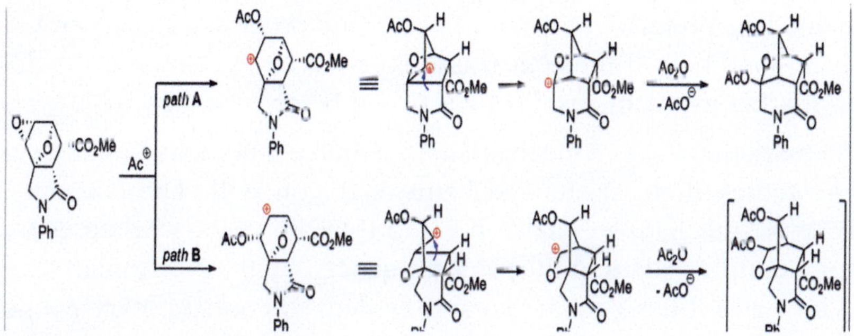

Figure 4.1. Plausible mechanism of Wagner.

Source: Image by Wikimedia Commons.

4.3. ON THE USE OF THE CONCEPT OF MECHANISMS IN THE SOCIAL SCIENCES

The interdisciplinarity of the mechanism approach is an intriguing feature. We may use modern biology as an illustration of this. According to Francis Crick, who won the Nobel Prize in 1962 for discovering the molecular structure of DNA, scientists in the twentieth century prefer to conceptualize in terms of "mechanisms" rather than "laws."

This is because the concept of "laws" is traditionally reserved for physics, which is the only field capable of producing explanations based on powerful and frequently contradictory principles with no notable exceptions. "What we see in life are mechanisms constructed with chemical components and frequently changed by other, later mechanisms added to the earlier ones."

The proportion of clearly stated mechanism-based explanations differs greatly between fields. Such kinds of explanations are rarely (if ever) employed overtly in history, although they are used rather frequently in economics and psychology. The concept of mechanism is particularly important in cognitive psychology.

"The information-processing approach [in cognitive psychology] assumes that perception and learning can be analyzed into a series of stages during which specific components ('mechanisms') perform certain transformations or recoding of the information coming into them," according to a well-known work.

Economists are frequently seen to think in terms of processes, in contrast to sociologists and historians, who are thought to be more concerned with social structures. "By economics–or, if you prefer, 'economics proper'–we signify the interpretative account of economic processes that operate inside any given condition of those institutions [covered by economic sociology], such as market mechanisms," argues Schumpeter.

The market is the one mechanism to which economists link the majority of their studies–their master mechanism, if you will. The concept of the market as a "mechanism" dates back to the 18th century when economics (via, for example, Adam Smith) was influenced by the Newtonian-Cartesian worldview, and it has become so self-evident to contemporary economists that the market is a mechanism that they frequently use the terms "market" and "market mechanism" interchangeably.

Much of twentieth-century neoclassical economics may be regarded as an attempt to describe progressively more parts of the economic process using market mechanisms: production, consumption, and distribution. As Tyler Cowen points out in his study of the usage of mechanisms in economics, the concept of mechanism is also inherent in the concept of equilibrium. It is worth emphasizing that economists' abilities to conceive in terms of mechanisms are often shown to non-economists only when they venture outside the usual confines of their subject.

Examples may be found in Albert Hirschman's Exit, Voice, and Loyalty (1970) and, to a lesser extent, in Thomas Schelling's Micro motives and Macro behavior. As economists steadily broadened their discipline's bounds to embrace a range of areas traditionally considered the realm of sociologists, such as the family and organizations, the distinction between the disciplines has grown to concern the sorts of theories utilized. One such distinction, although far from the only one, is the priority placed on explanatory processes.

4.3.1. The Use of Mechanisms in Sociology

As previously stated, sociology lags behind economics and many other sciences in terms of explicitly specified mechanism-based theories. The term "mechanism" is extremely widespread and has a long history in sociological studies, yet it is almost usually employed in a casual daily meaning. As an example of this tendency, consider its most likely first use in sociology.

Albion Small released General Sociology, a textbook on sociological theory, in 1905, and includes a list of the most essential sociological topics. Small provided examples such as "society," "social structure," "social status," and "social mechanism." Small, on the other hand, makes no meaningful attempt to explain the idea of social mechanism in the text of his work.

Small's definition of a social mechanism is utilized in a casual everyday fashion, as it is in modern sociology. This type of use is referred to as a "proto-concept" by Robert Merton, who defines it as "an early, rudimentary, particularized, and largely unexplained idea; a concept [on the other hand] is a general idea that, once defined, tagged, substantially generalized, and explicated, can effectively guide inquiry into seemingly diverse phenomena."

The term "mechanism" is rarely employed in sociological classics, even when the concept itself is there. Among the most well-known instances is the process at the heart of The Protestant Ethic, more specifically how austere Protestantism at one time in history led to changes in people's economic circumstances.

To summarize Weber's point, a believer's conversion to austere Protestantism began to place a religious emphasis on a specific sort of behavior, with the unintended result of creating a new norm for how to act in economic matters. Simmel's and Durkheim's writings both contain a number of key mechanisms. Simmel's usage of Tertius Gaudens is one example, as is Durkheim's examination of how the balance between individual and group influences the suicide rate. The term "mechanism" does not appear to have been used explicitly in sociology until after World War II. The most illuminating treatment of the subject, in our opinion, may be found in the writings of Robert Merton, who combined the concept of mechanism with the concept of middle-range theorizing.

Merton vehemently opposed any attempts to establish universal systems of sociological theory, instead advocating that sociological theory should focus on "social processes." Merton stated that the objective is to find a halfway ground between social regulations and description, and "mechanisms" are such a middle ground.

Merton defines social mechanisms as "social processes having designated consequences for designated parts of the social structure" in Social Theory and Social Structure, and argues that the main task of sociology is to "identify" mechanisms and determine under what conditions they "come into being," "fail to operate," and so on.

Merton briefly explains concrete methods that determine reference groups, generate dissonance, and express role-sets briefly. The most essential contribution of his work, in our opinion, is the idea of mechanisms as basic building blocks of middle-range theories. There has been little serious debate on mechanism-based theorizing in sociology since the fall of the Columbia School.

As far as we know, there is just one exception to a universal meta-theoretical debate within sociology, and that is a recent paper by Arthur Stinchcombe titled "The Conditions of Fruitfulness of Theorizing about Mechanisms in Social Science." Stinchcombe accurately observes in this piece that "we do not have a sufficiently supple arsenal of procedures for creating social science theory."

4.3.2. Mechanisms in the Social Sciences

A mechanism, like other scientific concepts, is part of many social scientists' everyday causal lexicon. Without specific theorizing about the nature of mechanisms, the term "mechanism" might refer to a cause, a causal route, or an explanation in this context. This informal and occasionally mechanical mode of communication is likely as ancient as the social sciences.

Certain negative connotations of the word "mechanical" meaning simplistic, inflexible, and reductionist thinking are also centuries old. Being part of the daily causal lexicon explains much of mechanistic language's intuitive appeal, while the negative implications have made some social scientists wary of the mechanistic shift. Mechanism theory has several origins in the social sciences. Rom Harré's philosophy of science is one of the most famous early sources.

Although his later work was prominent in social psychology, his philosophy of science had the most effective through the so-called critical realism movement. Roy Bhaskar is a significant theorist in this movement, and Harré's theories regarding causality and processes play an important part in his philosophy.

Regardless of Bhaskar's transcendental argument, Critical realism is still one of the most important metatheoretical trends in social science, owing to his layered explanation of reality and notions about essences and internal connections that have prompted philosophical misgivings and reservations about their applicability to the social sciences.

Although philosophers of mechanisms may be unfamiliar with critical realism, many social scientists believe that the concept of causal mechanisms is significantly related to this trend. Jon Elster was another early proponent of mechanism-based thinking, and his many works provide good illustrations of mechanistic thinking in action.

The early formulation, which states that a mechanism explains by establishing a continuous and contiguous chain of causal or deliberate ties between the explanans and the explanandum, is highly consistent with the broader mechanistic viewpoint. Although Elster's mechanisms are more psychological than social, his work has encouraged many social scientists to open up black boxes and reveal the gears and wheels of social processes' core machinery.

Elster has also firmly related the mechanistic attitude with the intellectual values of clarity and precision: he regards mechanism-based theorizing as clearheaded causal reasoning about social processes, and this is the foundation of the mechanistic approach for a huge group of social scientists. It is worth noting that there are few links between social scientists motivated by Bhaskar's critical realism and those who embrace Elster's approach.

In the social sciences, the concept of mechanism-based reasoning has several implications. In the 1990s, mechanistic thought began to gain traction. Social Mechanisms, edited by Peter Hedström and Richard Swedberg, was particularly significant in this regard. Many researchers who have been significant in the development of mechanism-based thinking in the social sciences have contributed to this collection.

Later, this method evolved into the analytical sociology movement, which was responsible for the most systematic development of the agenda of mechanism-based social science. Simultaneously, philosophers of the social sciences like as Daniel Little and Mario Bunge began to discuss mechanisms. Mechanisms became a focus of emphasis in political science much later.

Causal processes were essential to discussions about research techniques and causal inference. Especially important has been the development of process tracing methods for case-study research. In general, the disputes in political science and sociology have been relatively similar: supporters of mechanisms have challenged the oversimplified use of statistical techniques and the underestimation of the role of causal process beliefs in reasoning.

Although, there are some significant variances in the latter. In this section, first let's explore Coleman's diagram, which serves to identify the key problems of mechanism-based theorizing in the social sciences. It also provides a setting for social scientists to debate the attraction of mechanism-based explanations. In order to make sense of this discussion, shall distinguish between causal situations and causal mechanism schemes.

4.3.3. Coleman's Diagram

A diagram known as Coleman's boat is a good place to start when examining social mechanics. It is frequently employed in discussions of social processes, and one may argue that it has come to represent mechanism-based thinking in the social sciences. Coleman did not use mechanistic terminology, but the graphic represents mechanistic thinking in the social sciences in many respects.

The sociologist's preferred starting point for unpacking the figure is node D, which symbolizes the macrosocial explanandum. Node A denotes some macro-social variables related to D. Assume, for example, that A is the introduction of a government job-training program and D is the reduction in young unemployment.

Was the job-training program responsible for the drop-in unemployment? To put it another way, may arrow be understood as causal? The following is Coleman's major point. To substantiate causal assertions like these, it is crucial to understand how the proposed cause causes the outcome of the issue. This concept is made up of two parts. First of them is the substantiation of causal claims: statistical data on the relevant macro variables is typically insufficient to support such a claim. Coleman suggests testing the causal claim by determining if there is an experimentally accepted mechanism by which A causes D: if there is, the claim is supported; if not, it should be abandoned.

The second component is all about explaining. Coleman and other proponents of mechanism-based thinking emphasize the necessity of knowing how the impact occurred. As a result, even if it were feasible to demonstrate that variable A is a causal difference-maker for variable D, this would not serve as a theoretically adequate explanation: the explanation must specify forth.

The two concepts are essential to social-scientific disputes concerning causal processes. Let's go over both in depth later, but first, look at other parts of the diagram. The main idea in the graphic is that macro variables must be linked to agent behaviors. In most situations, these agents are human beings, but Coleman allows for various types of corporate agents to fill this position as well. Thus, the arrow depicts how changes in macro circumstances affect the relevant actors.

The change in A may cause changes in these agents' beliefs, wants, or other mental qualities, or it may affect the opportunities or incentives they face. These influences are referred to as situational mechanisms by Hedström and Swedberg. They discuss how social structures limit and facilitate people's options for action, as well as how cultural and social circumstances impact individuals' objectives, beliefs, habits, or cognitive frameworks.

The diagram's arrow represents the function of the theory of action in sociological explanation. The theory of action seeks to link changes in agents' opportunities and mental states to changes in their behaviors or acts. Coleman employed rational choice theory for this purpose, although other action theories might also be used.

The key point here is that mechanical explanations in the social sciences end at the level of individual action. Individual people's behaviors are the fundamental components of social processes, and social scientists do not seek explanations for these action-formation mechanisms: cognitive scientists and psychologists do.

Although both situational and action-formation mechanisms provide obstacles, Coleman contends that transformational mechanisms (arrow 3) are the most significant bottleneck in sociological theory.

Social scientists know quite a lot about how individuals' desires, beliefs, and possibilities are impacted by the social settings in that they are integrated (situational mechanisms), or how these desires, beliefs, and opportunities impact actions (action-transformation mechanisms), however when it tends to come to the connection between individual actions and social consequences, individuals are frequently forced to wave their hands.

Coleman's diagram was designed to emphasize this point: Individual-level descriptions seldom anticipate macro-level patterns, and the mechanism in which individual activities form societal patterns is rarely a straightforward process of aggregation. It is not an issue exclusive to individualistic theories: while holistic theories emphasize the contextuality and complexity of everything, their micro-macro assumptions are very simple in reality.

It is common to believe that macro facts simply reflect relevant micro facts and vice versa. Even in extremely simple situations, Schelling's well-known checkerboard model demonstrates how incorrect this assumption is: it demonstrates how inhabitants who do not want segregation can nonetheless end up in a highly segregated community.

Even from this simple model, it is evident that macro facts cannot be assumed to represent individual (or average person) choices. It is vital to comprehend how the macro result is influenced by the interconnected activities of people, or to comprehend "the rules of the game."

Coleman was never able to fully untangle the metaphor of the game's rules, but his discussion and examples piqued sociologists' interest in the theoretical issue of the micro-macro relationship. As a result, whereas others may talk vaguely about emerging macro-scale qualities, sociologists following Coleman take on the considerably more difficult task of describing how that emergence occurs.

The essential mechanical insight reflected in the figure is that understanding the relationship between macro-scale social facts requires understanding the entire chain of situational, action-formation, and transformational mechanisms. A mixture of these processes underpins each A-D link or causal relationship. Without these three, the explanation would not address all aspects of the micro-macro connection.

4.3.4. MECHANISM-BASED EXPLANATIONS

Mechanism-based views of explanations have been created in the social sciences, as well as in the philosophy of science, as alternatives to the once prevalent covering-law account of explanation.

While social scientists are familiar with famous counterexamples to that theory and philosophical problems raised by them (such as problems of explanatory relevance, the symmetry between explanation and prediction, asymmetry of explanation, and many problems in analyzing the concept of law), social scientists have been concerned primarily with the theory's evident improbability as a model for the social sciences.

In the social sciences, there are very few rules, and even those are best defined as explananda rather than explanations. Most non-truism social-scientific generalizations are fairly narrow in the area and have exceptions that tidy ceteris paribus criteria cannot encompass.

Furthermore, the technique of developing social-scientific theories in terms of axioms and laws has shown to be extremely ineffective, yielding incomplete collections of sterile generalizations plagued by unclear areas of application and high degrees of conceptual indeterminacy. Finally, for social scientists, the core concept of a covering-law theory is paradoxical.

The greatest social-scientific explanations appear to do more than merely subsume the phenomena under more general empirical regularities: a good explanation demonstrates how the proposed cause causes the effect to be described. This generative notion of explanation is at the heart of mechanistic ideas' initial attraction.

However, dissatisfaction with the covering-law theory has not resulted in widespread agreement on the concept of a mechanism. When people have diverse uses in mind, the same notion might be developed in many directions. As a result, the literature on social mechanisms is infamous for including seemingly contradictory definitions of mechanisms.

For example, Mahoney (2001) mentions 24, and later additions have increased the number. Certain critics (for example, Norkus, 2005) see multiplicity as a severe challenge for the mechanism-based explanation. Although the definitions are strictly conflicting, the majority of them could be regarded as efforts to capture the same underlying notions.

The absence of a commonly accepted definition for fundamental notions such as genes and species has not slowed the progress of biological sciences. Mechanism-oriented social scientists should agree without a general definition as long as they agree on the fundamental exemplars of mechanism-based explanations and have a comparable grasp of their basic features. Thus, rather than providing a comprehensive definition of mechanism-based explanations, the major problem is to get agreement on archetypal examples of mechanistic explanations.

4.3.5. Agent-Based Simulation and Generative Explanation

Among the most important aspects of mechanism-based explanation is that it should indicate how the features, actions, and relationships of components cause the phenomena to be described (Glennan & Darden, this volume).

Because of this interest in generating processes, generative sufficiency is a fundamental issue in the evaluation of explanations: the proposed explanation should, at the very least, be capable of producing the outcome under the given conditions. It is, nevertheless, difficult to demonstrate generative sufficiency.

The social consequences of interest are often the product of many people acting and communicating with each other over long stretches of time. Provides the most powerful account-based marketing (ABM)-based description of the concept of generative explanation. "If you didn't grow it, you didn't explain it," Epstein says. He meant that obtaining the macro-level outcome with ABM is a required prerequisite for explaining it.

It may be argued, however, that just "expanding" the phenomena of interest is insufficient to generate a thorough understanding. The key problem is to figure out how the stated micro configuration causes the occurrence. Thus, Macy and Flache highlight an important point in reformulating Epstein's slogan: "If you don't know how you grew it, you didn't explain it." In any event, social scientists employing ABM are dissatisfied with mechanical explanations as simple explanations.

4.4. THE EXPLANATORY IMPORTANCE OF SOCIAL MECHANISMS

The main argument of this chapter is that identifying and analyzing social processes is critical for the advancement of social science theory and research. But what precisely is a mechanism, and why should we pay attention to mechanisms rather than statistical connections or other types of linkages between the things of interest?

Providing a clear yet sufficiently wide description of a social mechanism that captures the core of the notion is far from straightforward. According to Harre, one essential distinguishing feature of an explanatory mechanism is the role it serves in an explanatory account. Assume we've discovered a consistent link between two things, say / and O. To describe the link between them, we look for a mechanism, M, that when the cause or input, /, occurs, it causes the effect or result, O. The quest for mechanisms implies that we are not content with just finding systematic covariation between variables or occurrences; a good explanation demands that we be able to explain the social "cogs and wheels" that have brought the connection into existence.

A mechanism, as Scheling underlines, may be thought of as a systematic group of assertions that offer a convincing account of how / and O are related to one another. This perspective on probable causes differs significantly from the standard covering-law model championed by Carl Hempel and his supporters.

An explanation of an occurrence, according to Hempel, requires subsuming the event under general law. A proper explanation must thus clarify the generally applicable law as well as the factors that make the law appropriate in the individual situation. Deterministic rules, according to Hempel, are extremely rare in the social and historical sciences.

Instead, the "rules" that might be invoked in the social sciences are probabilistic in character (i.e., they state that the occurrence of a particular event will come about with such and such probability if certain specified conditions are at hand). Because this type of explanation includes merely when applying general law to a given circumstance, the insights gained are directly proportionate to the depth and strength of the "probabilistic law."

Whether this law is just a statistical association, as Hempel claims is the norm inside the social and historical sciences, the particular reasoning will provide no more additional insight than that of the law therefore it will generally just strongly recommend that a connection is likely to be present while providing no clue as to why this is likely to be the case.

For these reasons, we agreed with von Wright that it is best "not to state that the inductive probabilistic model [of Hempel] explains what happens, but instead to say that this really justified certain expectations and forecasts." Because it does not require that the mechanism linking explanans be specified, the covering-law model justifies the use of "blackbox" explanations in the social sciences.

In order to have an acceptable explanation, an explanandum must be stated. This omission has allowed for sloppy research, and one of the key benefits of the mechanism-based approach is that it enables (or promotes) deeper, more direct, and finer-grained explanations. As a result, the quest for generative processes aids in distinguishing between actual causation and fortuitous connection, and it broadens our knowledge of why we perceive what we experience. The recent dispute over the potential health consequences of electromagnetic fields exemplifies the significance that the quest for mechanisms plays in discriminating between spurious and actual correlations.

Some epidemiological research has discovered an empirical link between electromagnetic field exposure and pediatric leukemia. Nevertheless, the significance of these empirical findings is diminished by the fact that no known biological mechanism can explain how low-frequency magnetic fields might conceivably cause cancer.

Bennett believes it is even more improbable that a mechanism would ever be discovered since such a mechanism would have to contravene well-established basic laws. The absence of a viable mechanism raises the possibility that the weak and fairly unsystematic empirical evidence described in this epidemiological literature represents unmeasured confounding variables rather than a real causal connection.

4.4.1. Variables versus Social Mechanisms

The extensive use and understanding of survey analysis, as well as the statistical tools required for interpreting such data, has undoubtedly increased sociologists' abilities to characterized social situations and evaluate sociological hypotheses.

However, as Srensen emphasizes in, the rising use of these tools has facilitated the emergence of a variable entered kind of theorizing that pays little regard to explanatory mechanisms. Coleman correctly identified this kind of sociology as "individualistic behaviorism."

The guiding concept underlying this form of reasoning, known as "causal modeling," is the idea that numerous individual and environmental "determinants" may and should explain individual behavior, and the goal of the analysis is to estimate the causal effect of the various variables reflecting these determinants.

This emphasis on "causal" explanations of behavior, according to Coleman, represented a significant shift from the sort of informational account used throughout the previous tradition of community studies: "Each way to describe this transformation is to suggest that statistical association in both variables has largely been replaced significant link between events as the basic tool of analysis and interpretation."

Variables, not actors, perform the actions in the causal-modeling tradition. In the mid-1970s, an interaction between Robert Hauser and Raymond Boudon highlighted the contradiction between a variable-centered causal approach to sociological theory and a generative view emphasizing the relevance of social processes.

The setting for this conversation was Hauser's assessment of Boudon's book on education and inequality. Boudon constructed a theoretical model in this book that he anticipated would explain a number of seeming inconsistencies observed by empirical studies on social mobility.

4.5. SOCIAL MECHANISMS: SOME SELECTED EXAMPLES

To clarify and define the idea of a general social mechanism underlying a variety of different social phenomena, we will briefly examine three well-known sociological theories–the self-fulfilling prophecy (Robert Merton), network diffusion (James Coleman), and threshold-based behavior (Mark Granovetter)–and argue that they all rely on the same basic belief-formation mechanism.

The self-fulfilling prophecy is one of the most well-known mechanism-based hypotheses in sociology, having been proposed by Robert Merton in a major piece in 1948. The essential assumption is that a misleading initial characterization of a situation elicits action that gradually leads to the erroneous conception being real Merton's primary example for demonstrating his point is a bank run.

If a rumor of insolvency spreads, some depositors will remove their funds. Their withdrawal will fuel the rumor, partially because the withdrawals may harm the bank's financial condition, but more crucially because the act of withdrawing indicates to others that something is genuinely wrong with the bank. This leads to even more withdrawals, which undermines faith in the bank even further, and so on.

Because of how this system works, even a seemingly healthy bank might become bankrupt if enough depositors remove their funds at the first mistaken impression that the bank is bankrupt. The study of network dissemination mechanisms is a thriving topic of sociological research at the moment; Strang and Tuma, Hedstrom.

This line of research has been heavily influenced by Coleman, Katz, and Menzel's famous investigation of the dissemination of new medicine. Their key discovery was that physicians' standing in various professional networks impacted the dissemination process, particularly in the time immediately after the introduction of new medicine to the market.

Granovetter's threshold theory of collective behavior is our final illustration. According to Granovetter, an individual's decision to join in collective behavior is frequently influenced by how many other actors have already opted to participate.

He went on to explain that actors differ in terms of the number of other actors who must participate before they decide to do the same, and he coined the word "threshold" to represent this individual heterogeneity. An actor's threshold defines the proportion of the group that must have joined before the actor in question is prepared to do so, and Granovetter's study revealed that even little changes in thresholds can result in radically different collective outcomes.

4.6. SOCIAL MECHANISMS: A TYPOLOGY

As various writers in this book point out, explaining most real social events or situations necessitates the use of several simple processes; one is insufficient. Occasionally these systems operate against one another, and sometimes they complement one another. In any case, the diversity of processes necessitates the development of a typology that categorizes them in a meaningful way.

The one we will offer here departs from James Coleman's (1986) well-known paradigm for conceptualizing collective social action, known as the macro-micro-macro model. The basic premise of this paradigm is that effective explanations of change and variation at the macro level include demonstrating how macro states at one moment in time impact the conduct of individual actors, and how these actions produce new macro states at a later time.

That is, rather than analyzing relationships between phenomena solely on the macro level, one should always attempt to establish how macro-level events or conditions affect the individual (Step 1), how the individual assimilates the impact of these macro-level events (Step 2), and how a group of individuals, through their actions and interactions, generate macro-level outcomes (Step 3).

This approach to thinking about social activity lends itself quite well to a typology of mechanisms: There are macro-micro mechanisms, micro-macro mechanisms, and micro-macro mechanisms, and each will be discussed briefly. In the first two categories of scenarios, the actor is a single person, and the mechanism is internal (and hence "psychological" or "social-psychological"); in the third type, there are several actors, and the mechanism is often external (and "social").

The following definition of a mechanism contains the core of the first two sorts of mechanisms, which focus on discrete parts of behavior: "Mechanisms are hypothetical causal models that explain individual behavior and have the form 'given specific conditions K, an agent will perform x because of [mechanism].' M with a chance Stinchcombe's earlier-mentioned concept of mechanisms as mediating between lower and higher levels, on the other hand, focuses more directly on social interaction and the effects of social activity.

The first of the three sorts of mechanisms cover the macro-to-micro transition, in Coleman's language, and we'll name it a situational mechanism after Stinchcombe's notion. The individual actor is exposed to a certain social context, which affects him or her in a specific way. These kinds of processes are at the heart of Erving Goffman's work on public conduct and Karl Popper's style of situational analysis.

The previously discussed belief-formation mechanism, opportunity-generating mechanisms such as White's vacancy chains, and preference-formation mechanisms such as those expressed in the concept of reference groups are prototypical examples of general social mechanisms that link a social structure or other macro-sociological event or states about a particular actor's ideas, wants, and opportunities second sort of mechanism, known as an action-formation mechanism, will be found at the micro level.

This method demonstrates how a specific mix of human goals, beliefs, and action possibilities results in a certain action. At this level, a variety of psychological and social-psychological systems are at work. Different forms of action processes are shown by general choice theories as well as more particular theories such as Leon Festinger's theory of cognitive dissonance and George Ainslie's on discounting.

The third sort of mechanism, which we propose to term a transformational mechanism, deals with the micro-to-macro transition. A number of individuals engage with one another in this scenario, and the precise process (which varies depending on the nature of the contact) demonstrates how these individual activities are changed into some type of collective effect, whether intentional or accidental.

Several other theories discussed in this book, including Schelling's tipping model, basic game-theoretic models like the tragedy of the commons, and neoclassical market models, are instances of transformational mechanisms.

4.7. SOCIAL MECHANISMS AND SOCIAL DYNAMICS

The role, or exclusive role, of "methodological individualism," the assumption that the ultimate unit of analysis is a rational, or at least purposeful, individual, is widespread concern for social phenomena. For some of the authors here, any social phenomenon that cannot be reduced to individual conduct is a black box, and hence unsatisfying.

Some people believe that what's within a black box is a social mechanism or a collection of social processes. But what exactly are social processes, and where do they come in? And are social mechanisms small, large, or extremely large? Was Keynesian theory a social mechanism? Was the weapons race a social mechanism? Was inflation a social mechanism?

Or is giggling Such a mechanism, or yawning, or rumor spreading? In terms of its relationship between social mechanisms and theories, A suggestion, that a theory may include many social mechanisms, but a social mechanism may also include many theories. A specific question that emerges is whether a social process may be completely mathematical.

4.8. SOME ELEMENTARY MECHANISMS

4.8.1. Adaptive Preferences versus Wishful Thinking

According to Festinger's theory of cognitive dissonance, discord occurs when a person has two or more "cognitions" that contradict one another. In this context, cognitions encompass not just regular factual ideas, but also consciously held values and mental representations of the subject's decisions or activities.

Inconsistency is founded on "expectations about what goes with what built up through time, including conceptions of logical relations, cultural mores, and learnt empirical connections between experiences." Thus, if a guy has just purchased an automobile of brand X, it is expected that he will not perceive brand Y to be superior.

To use another well-known paradigm of cognitive dissonance research, if a subject is requested to write an essay arguing for abortion and decides to do so, even if the benefits are minor, expectations are dashed when it is discovered that he is truly vehemently opposed to abortion. Dissonance reduction (or avoidance) occurs through altering or suppressing some of the discordant cognitions, as well as occasionally introducing new ones.

Despite certain ambiguities in Festinger's initial statements, the process must be regarded as unconscious. In the vehicle example, dissonance may be avoided by seeing commercials for the car one has recently purchased and ignoring ads for other manufacturers. These actions cannot be the result of intentional choices made to reduce dissonance.

Because no bolstering could occur if it was known that the advertising was read or ignored in order to boost one's confidence in the decision one had already taken. One "simply gravitates" toward acts that confirm the correctness of the decision. In the essay-writing case, the dissonance-reducing modification is more plainly constrained to be unconscious.

When respondents are unable to convince themselves that their actions are justified by a lack of choice or a large payoff, they lessen their dissonance by developing a more positive attitude toward the viewpoint they advocated (Figure 4.2).

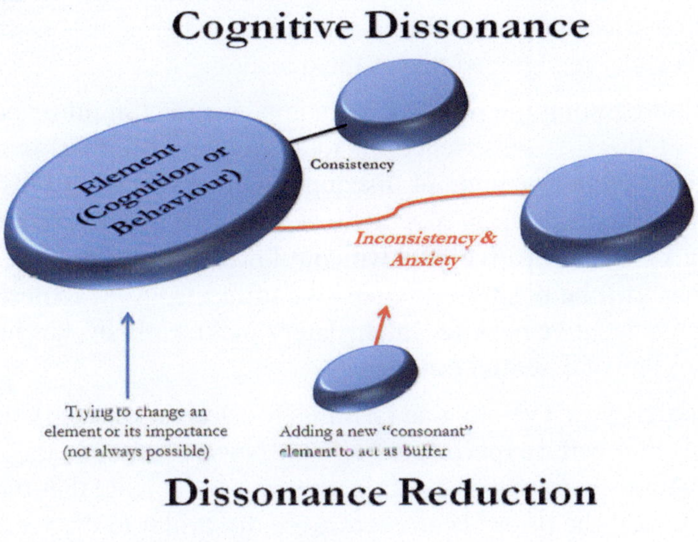

Figure 4.2. Cognitive dissonance diagram.

Source: Image by Wikimedia Commons.

4.8.2. Contrast Effect versus Endowment Effect

Amos Tversky proposed (personal communication) in the mid-1980s that previous experience had a twofold influence on present wellbeing. On the one hand, there is an endowment effect: a good memory is a good memory, and a bad memory is a terrible memory. As a result, a good past tends to improve the present, whereas a negative past tends to worsen it. There is, however, a contrast effect: A positive experience in the past tends to discount fewer positive events in the present, and a negative incident in the past will similarly cast the present in a favorable light. Dinner at a fantastically wonderful French restaurant may lead to further meals at French restaurants (and perhaps at other restaurants, too).

Less than one would have done otherwise Recovery from disease, on the other hand, makes you appreciate a regular level of health. We may pose various questions based on the existence and regular operation of these two systems. First, there is a type B2 mechanism: Will the overall effect be positive or negative if a positive or bad experience causes a negative or positive contrast effect (assuming no endowment effect)?

This has been a hotly debated topic since it was realized that "if the best can occur only seldom, it is preferable not to include it in the spectrum of experiences at all." Second, there is a type B mechanism. What is the net impact of contrast and endowment, of previous experiences on welfare later in time? Tversky recognized this as the question.

Third, integrating the first two questions, we may inquire about the net influence of the first experience on total wellbeing, either as discounted to a previous period or without discounting. The net effect on subsequent welfare may be negative in a specific situation (a negative contrast effect being larger than a positive endowment effect), but the net effect on total welfare may still be positive (the positive utility from the experience itself offsetting the negative net effect at the later times). Nobody, has investigated the third and most essential question.

The majority of Tversky and Griffin's research is devoted to examining net impacts in specific experimental circumstances. Although they claim that their predictions were validated, this turns out to indicate that there was no contrast effect if the prior occurrences were dissimilar to the present ones.

Furthermore, they highlight that the loss aversion principle implies (and was proven) that the negative contrast impact following a high payout will be bigger than the positive contrast effect following a low payoff. They do not, however, provide any previous reasons for expecting that the contrast effect will outweigh the endowment effect or vice versa when both are in action.

It comes out that the endowment effect was stronger in one of their two tests, whereas the two effects were nearly equal in the other. Although loss aversion is mentioned as a possible reason for the difference between the two trials, no explanation is provided for the findings achieved in any of them. Later, George Loewenstein and Amos Tversky applied Tversky's notion to a broader range of experiences.

In addition to endowment and contrast effects arising from one's own previous experiences, we discovered analogous effects arising from anticipating one's future experiences, other people's experiences, and just imagined or counterfactual experiences. Because the phrase "endowment" does not apply in these other instances, we adopted the wider term "consumption impact."

It also addressed the issue of net effect to some extent that there is a change from a dominating consumption impact to a dominant contrast effect in interpersonal comparisons at the level of equality (Figure 4.3).

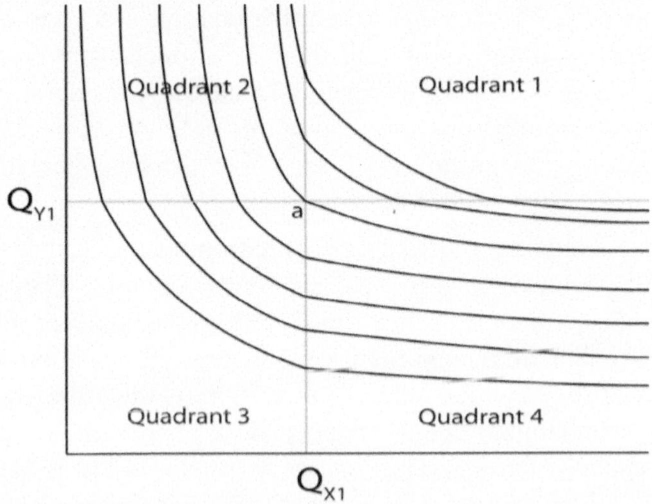

Figure 4.3. The endowment effects divide the plane into 4 quadrants.

Source: Image by Wikimedia Commons.

4.9. ELEMENTS OF SOCIAL MECHANISMS

Mechanisms in social science are made up of two types of abstract parts. The initial construction is the actor specification (i.e., determining which type of man or Humanoid is to be assumed) (i.e., the casting of the mechanism by stand-ins for humans).

The second set is the presumed structure or sort of setting in which the Humanoids work (i.e., the staging of the mechanism by the guide rails along which the actors can move.) The goal of building the mechanism is to determine its systemic effects (i.e., the plotting–what happens to the actors or to the structure as a consequence of their combination). It is precisely this mix of factors actor assumptions, structural assumptions, and so on–that makes and how they produce a logical dynamic which separates a social mechanism from other phenomena like covering laws, simple correlations, and so on.

4.10. CONCLUSION

In this chapter, the inquiry is, whether mechanisms can perform this service and, if so, how. The case on behalf of mechanisms is sometimes posed in terms of the claim that reliable causal inference in social science is impossible without knowledge of mechanisms. On the positive side, we can infer that X is a cause of Y if we know that there is a mechanism through which X influences Y. It is sometimes claimed that the consideration of social mechanisms can significantly ameliorate this problem. Kincaid argues that if no plausible mechanism running from X to Y can be conceived of, then it is safe to conclude that X does not cause Y.

As explained, neither of these lines of reasoning entails that mechanisms are necessary for causal inference in social science. They argue that knowledge of social norms and practices can be attained in a manner that is independent of the problem of unmeasured common causes. The effectiveness of the negative side of the account is undermined by the ease of imagining mechanisms connecting nearly any two variables representing aspects of social phenomena. The positive account is not helpful unless some explanation is given of how it is possible to learn about social mechanisms without running afoul of the problem of unmeasured common causes.

REFERENCES

1. Hedstrom, P., & Swedberg, R., (1998). *Social Mechanisms an Analytical Approach to Social Theory*. [e-Book] https://pedropeixotoferreira.files.wordpress.com/. Available at: https://pedropeixotoferreira.files.wordpress.com/2014/03/peter-hedstrocc88m-richard-swedberg-social-mechanisms_-a.pdf (accessed on 27 June 2022).
2. Prügl, E., (2017). *Social Mechanisms: The Art of World-Making*, 160–174. [online] Available at: https://www.taylorfrancis.com/chapters/edit/10.4324/9781315268934-25/social-mechanisms-elisabeth-pr%C3%BCgl (accessed on 27 June 2022).
3. Thagard, P., (2019). Social mechanisms. *Mind-Society*, 48–76.
4. Ylikoski, P., (2017). *Social Mechanisms.* [online] Available at: https://www.researchgate.net/search/publication?q=Social%20Mechanisms (accessed on 27 June 2022).
5. Ylikoski, P., (n.d). *Social Mechanisms*. [online] Academia.edu. Available at: https://www.academia.edu/26763898/Social_Mechanisms (accessed on 27 June 2022).

CHAPTER 5

Development and Modification of Social Behavior

CONTENTS

In this chapter, the development and modification of social behavior is discussed. It covers what behavior modification is, how it started, what its techniques are and what is its significance. This research is significant in establishing how Cognitive behavioral therapy has been used to treat multiple disorders.

Some ideas by Lewin, also known as father of social psychology have been addressed, including his person-situation interaction. It then talks about social psychology, its history, social cognition, social neuroscience and social influence. It discusses the ideas of Evolutionary Adaptation, Human Characteristics and Self-Concern. It also explores social norms, culture, individualism, and collectivism.

5.1. BEHAVIOR MODIFICATION

The tactics used to try to reduce or improve a specific type of behavior or reaction are known as behavior modification. This may appear to be a pretty technical term, but we all use it on a regular basis. This is how parents teach their children the difference between good and wrong. Therapists utilize it to encourage their patients to adopt healthy habits.

It is used by animal trainers to teach a pet and its owner to obey each other. We use it in our friendships and romantic interactions as well. Our reactions to them teach them what we enjoy and dislike. Conditioning is at the heart of behavior modification. Conditioning is a type of education. Classical conditioning and operant conditioning are the two main types of conditioning (Figure 5.1).

Figure 5.1. Behavior modification.

A specific stimulus or signal is used in classical conditioning. If a family member came to the kitchen every time you prepared cookies because of the delightful smell that would be an example of this. The second form is operant conditioning, which involves the use of a reward and punishment system. When a dog obeys an order, dog trainers frequently utilize this technique to reward them with a special goodie.

These beliefs gave rise to behavior modification, which reinforced the idea that just as behaviors can be acquired, they can also be unlearned. As a result, a variety of approaches for eliciting or inhibiting behavior have been created. This is how the concept of behavior modification was born.

5.1.1. Techniques of Behavior Modification

The goal of behavior modification is not to figure out why or how a certain behavior began. Instead, it focuses solely on modifying behavior, and various strategies are employed to do so. This comprises:

- Positive reinforcement
- Negative reinforcement
- Punishment
- Flooding
- Systematic desensitization
- Aversion therapy
- Extinction

A positive stimulus is associated with a positive response. Teachers that use stickers to reward their students for good grades are a fantastic illustration of this.

Negative reinforcement occurs when a behavior is paired with the absence of a negative stimulus. A good example would be a child who throws a tantrum because he or she refuses to eat veggies and then has their vegetables taken away.

Punishment is used to weaken habits by associating them with an unpleasant stimulus. Detention as a result of bad behavior is one example of punishment.

Flooding entails repeatedly and intensely exposing people to fear-inducing objects or situations. Flooding would be forcing someone who is afraid of snakes to hold one for 10 minutes. To treat phobias, systematic desensitization is utilized, which entails educating a client to remain calm while focusing on their concerns. Someone who has a strong phobia of bridges, for example, would begin by looking at a photo of a bridge, then imagine themselves standing on a bridge, and finally walk across an actual bridge.

> Aversive Therapy entails nurturing good thoughts about a person while creating love for them. Instead, than pointing out a person's flaws, one should focus on the positive qualities that person possesses, and one should not be averse to others. Aversion is a boomerang that attacks the person who launched it. As a result, aversion should be avoided.

Extinction entails forcing a person to abandon undesirable behavior. When a child is placed in a timeout for undesirable behavior, that behavior may eventually end.

5.1.2. Significance of the Study

Many people have criticized the concept of punishment in recent years; however, these concerns usually refer to the addition of some disagreeable event rather than to negative punishment (time-outs). Positive punishment is only used by board certified behavior analysts in extreme cases when all other forms of treatment have failed and the behavior to be changed poses a risk to the person or others.

Positive punishment in clinical settings is mainly limited to employing a spray bottle filled with water as an aversive event. When handled incorrectly, more aversive punishment can result in affective (emotional) disorders, as well as the recipient of the punishment desiring to avoid it (i.e., "not be caught"). Many residential treatment center programs include behavior modification programs.

They have demonstrated success in lowering recidivism in both youth and adult criminals. Compliments, acceptance, encouragement, and affirmation are one approach to provide positive reinforcement in behavior modification; a ratio of five compliments for every one complaint is typically viewed as beneficial in changing behavior and even generating stable marriages.

The correct behavioral intervention can have a big impact on the whole system. Parent management training programs, also known as behavioral parent training programs, have demonstrated to be somewhat cost-effective in the treatment of conduct disorder. As a result, such intervention can have a significant impact on socializing the child while also assisting the parent in escaping poverty.

Persons who perform behavioral engineering search for and value this level of influence, and results like these have led the Association for Behavior Analysis International to declare that those receiving therapies have a right to effective therapy and a right to effective education.

Attention deficits, impulsivity, anger, depression, noncompliance, attributions, motivation, social skills, and meta-cognition, as well as scholastic deficiencies in reading, written expression, handwriting, math, and spelling, have all been addressed with cognitive behavioral therapies. In this light, the current research is significant.

5.2. SOCIAL PSYCHOLOGY

The discipline of social psychology is fast expanding and has a rising influence on how we think about human behavior. The findings of social psychologists are frequently reported in newspapers, journals, websites, and other media, and the outcomes of social psychological research are affecting decisions in a wide range of domains. Let's start with a brief overview of social psychology's history before moving on to a discussion of the science's fundamental ideas.

5.2.1. The History of Social Psychology

When scientists first began to deliberately and formally measure human thoughts, feelings, and behaviors, the field of social psychology was born (Kruglanski & Stroebe, 2011). Before 1900, the first social psychology group behavior experiments were undertaken (Triplett, 1898), and the first social psychology textbooks were published in 1908 (McDougall, 1908/2003; Ross, 1908/1974).

Kurt Lewin and Leon Feininger, social psychologists, expanded the experimental technique to analyzing behavior in the 1940s and 1950s, establishing social psychology as a rigorous scientific subject. Because he pioneered many of the discipline's key concepts, such as an emphasis on people's dynamic relationships, Lewin is sometimes referred to as "the father of social psychology."

Festinger and other social psychologists published Research Methods in the Behavioral Sciences in 1954, in which they emphasized the need of measuring variables and using laboratory experiments to systematically verify research hypotheses regarding social behavior. He also mentioned that in some tests, it might be essential to deceive the participants about the genuine nature of the study.

Researchers attempting to understand how Adolf Hitler, the German dictator, could have instilled such strong allegiance and horrifying actions in his people during World War II utilized social psychology. Muzafir Sherif (1936) and Solomon Asch (1952) studies on conformity, as well as Stanley Milgram (1974) studies on obedience, demonstrated the relevance of conformity demands in social groupings and how persons in authority might produce obedience, even to the point of causing serious harm to others.

In his well-known "prison study" (Haney, Banks, & Zimbardo, 1973), Philip Zimbardo discovered that the interactions of male college students hired to play guards and prisoners in a simulated jail were so violent that the study had to be stopped early (Figure 5.2).

Figure 5.2. Social psychology.

Other areas were swiftly added to the study of social psychology. Leonard Berkowitz (1974) pioneered the research of human hostility, and John Darley and Bibb Latané (1968) established a model that helped explain when people do and do not help those in need.

Other social psychologists, such as Irving Janis (1972), focused on group behavior, attempting to understand why clever people sometimes made decisions that resulted in bad outcomes while working together. Other social psychologists, such as Gordon Alport and Muzaffar Sheriff, focused on intergroup relations in order to better understand and maybe reduce stereotyping, prejudice, and discrimination.

Social psychologists testified as expert witnesses in the 1954 U. S. Supreme Court case Brown v. Board of Education, which helped eliminate racial segregation in American public schools. Social psychologists continue to testify as expert witnesses on these and other matters (Fiske, Bersoff, Borgida, Deaux, & Heilman, 1991).

In recent years, social psychology ideas have even been used to construct anti-violence programs in nations that have undergone genocide (Staub, Pearlman, & Bilali, 2010).

In the latter half of the twentieth century, social psychology expanded into the study of attitudes, with a focus on cognitive processes. Social psychologists developed the first formal models of persuasion during this time, with the objective of understanding how advertisers and others could communicate their ideas in the most effective way possible (Eagly & Chaiken, 1993; Hovland, Janis, & Kelley, 1963).

The cognitive processes that people utilize while evaluating messages and the relationship between attitudes and behavior were the focus of these approaches to attitudes. During this time, Leon Festinger established his influential cognitive dissonance theory, which became a basis for subsequent study (Festinger, 1957).

In the 1970s and 1980s, social psychology grew even more cognitive, as researchers exploited breakthroughs in cognitive psychology to inform the field, which were mostly based on advances in computer technology (Fiske & Taylor, 2008).

Social cognition was the focus of these scholars, which included Alice Eagly, Susan Fiske, E. Tory Higgins, Richard Nisbett, Lee Ross, Shelley Taylor, and many more.

Social Cognition—an understanding of how our knowledge about our social worlds develops through experience and the influence of these knowledge structures on memory, information processing, attitudes, and judgment. Furthermore, the extent to which both cognitive and motivational processes can skew human decision-making was revealed (Kahneman, Slovic, & Tversky, 1982).

The study of social psychology has expanded even further in the twenty-first century. Interest in how social situations influence our health and happiness, the importance of evolutionary experiences and cultures on our behavior, and the field of *social neuroscience*—the study of how our social behavior both influences and is influenced by the activities of our brai*n* (Lieberman, 2010). Social psychologists are always looking for new techniques to assess and understand social behavior, and the discipline is always changing. We can't foresee where social psychology will go in the future, but we can be confident that it will remain alive and well.

5.2.2. The Person and the Social Situation

The study of the dynamic relationship between individuals and others around them is known as social psychology. Each of us is unique, and our personal attributes, such as personality traits, desires, motives, and emotions, have a significant impact on our social conduct.

However, the social situation—the people with whom we contact every day—has a significant impact on our conduct. Friends and relatives, classmates, religious groups, people we see on TV, read about, or connect with online, as well as those we think about, recall, or even imagine, are among these people.

Human conduct, according to social psychologists, is determined by both a person's traits and the social setting. They also feel that a person's social setting has a greater impact on conduct than his or her own attributes.

The study of social situations is at the heart of social psychology. Social influence is the process by which other people impact our thoughts, feelings, and behaviors, as well as the process by which we influence theirs.

Perhaps you can understand how social influence influenced Raoul Wallenberg's decisions and how he, in turn, influenced others.

In an essential equation, Kurt Lewin codified the simultaneous influence of person variables and situational variables, known as the person-situation interaction:

Behavior = f (person, social situation).

The conduct of a given person at any given time is a function of (depends on) both the person's qualities and the effect of the social setting, according to Lewin's equation.

5.2.3. Evolutionary Adaptation and Human Characteristics

Person refers to the attributes of an individual human being in Lewin's equation. People are born with abilities that enable them to interact successfully with others in their social environment. New-born's can recognize faces and respond to human voices, young children acquire language and form connections with other children, adolescents have a sexual curiosity and are destined to fall in love, most adults marry and have children, and most people get along.

People share these features since we've all been molded in the same way through human evolution. Humans have unique social abilities as a result of our genetic code, which is essential for survival.

The ability to engage in social behaviors, like good eyesight, physical strength, and disease resistance, helped our forefathers survive. We are quick to cast judgment on others, to aid those in need, and to enjoy working together in social groupings because these characteristics helped our forefathers adapt and were passed down the generations (Ackerman & Kenrick, 2008; Barrett & Kurzban, 2006; Pinker, 2002).

Our exceptional social abilities are owed in great part to our huge brains and the social intelligence they bring (Herrmann, Call, Hernández-Lloreda, Hare, & Tomasello, 2007).

Evolutionary adaptation is the belief that our nature, including most of our social behavior, is mainly determined by our evolutionary background (Buss & Kenrick, 1998; Workman & Reader, 2008). In evolutionary theory, fitness refers to how much a given quality aids an individual organism in surviving and reproducing at a higher rate than other members of the species who lack the characteristic.

Fitter animals pass on their genes more successfully to subsequent generations, making fitness-related features more likely to become ingrained in the organism's nature than non-fitness-related characteristics. It has been suggested, for example, that the emotion of envy has persisted in men over time because men who experience jealousy are more physically fit than men who do not.

According to this theory, jealousy causes males to defend their partners and guard against competitors, increasing their reproductive success (Buss, 2000).

Although our biological composition helps us to become human, it's crucial to remember that our genes don't really decide who we are. Genes, on the other hand, give us our human qualities, which give us the predisposition to behave in a "human" manner. Despite this, each human being is distinct from the others.

Evolutionary adaptation has given us two main incentives that guide and assist us in leading productive and effective lives. One of these motives is concerned with the self—the want to defend and enhance oneself and those mentally close to us; the other is concerned with the social situation—the desire to identify with, accept, and be accepted by others. These two motivations will be referred to as self-concern and other-concern, respectively.

5.2.4. Self-Concern

The urge to safeguard and promote our own life and the lives of those close to us is the most basic propensity of all living organisms, and the focus of the first human motivation. Humans are driven to seek out food and water, build suitable shelter, and defend themselves from harm. This is vital because we can only survive if we can achieve these basic objectives.

We do the same for our relatives—those who are genetically linked to us—out of a desire to maintain and strengthen the self. Kin selection is a strategy used by humans and other animals to prefer the reproductive success of relatives, often at the expense of the individual's own survival. Kin selection occurs because behaviors that improve the fitness of relatives, even if they impair the fitness of the individual, may benefit the survival of the group as a whole, according to evolutionary principles.

We want to safeguard, improve, and increase the well-being of our **in-group**, which includes those we consider to be similar and significant to us and with whom we have close social ties, even if they do not share our DNA.

Perhaps you recall assisting friends with moving all of their belongings into a new home when you would have wanted to be doing something more beneficial to yourself, such as studying or relaxing.

You wouldn't have done anything like this for strangers, but you did it for your buddies because you cared about them. The desire to support people we care about, even if they are not connected to us, is likely owing in part to our evolutionary background, when the people we cared about were frequently our relatives.

5.2.5. Other-Concern

Although we are primarily concerned with the survival of ourselves, our family, and those we consider to be like and significant to us, we also want to connect with and be accepted by others in general—the goal of other-concern. We live in communities, work in work groups, worship in religious groups, and participate in sports teams and clubs together. Joining forces with others, even strangers, helps us achieve a core goal: finding a loving partner with whom we can start a family.

Our social relationships can give us with opportunities that we would not otherwise have. We can go to the shop and buy milk or eggs, or we can engage a carpenter to construct us a home. And we supply commodities and services to others through our own work.

This mutual cooperation is advantageous to both us and those around us. We also join because we love being around other people, being a member of social groups, and contributing to social debate (Leary & Cox, 2008).

The other-concern motive implies that we do not always priorities ourselves. Being human also entails caring for, assisting, and collaborating with others. Despite the fact that our genes are "selfish" (Dawkins, 2006), individuals are not necessarily so.

Helping others, even those who are not related to us, may promote the survival of our own genes (Krebs, 2008; Park, Schaller, & Van Vugt, 2008). Humans participate in altruistic acts in which they aid others, often at a cost to themselves, just as birds and other animals may put out alarm sounds to other creatures to notify when a predator is nearby.

In short, human beings act morally toward others because they recognize that harming others without a compelling reason is wrong, and they show compassion and even charity toward others (Goetz, Keltner, & Simon-Thomas, 2010; Turiel, 1983).

Negative behaviors toward others, such as bullying, cheating, stealing, and aggressiveness, are as a result odd, unexpected, and socially unacceptable. Of course, this does not imply that individuals are always kind, helpful, or nice to one another; powerful social contexts can and do produce undesirable actions. However, because of the inherent human impulse of other-concern, hatred, and violence are the exception rather than the rule.

The goals of self-concern and other-concern can sometimes overlap. When one falls in love with another person, it is partly motivated by a want to connect with another person, but it is also motivated by a desire to feel good about themselves. And one volunteers to help people in need not only for their benefit but also for their own.

When one assists others, they feel wonderful. However, there are moments when the goals of self-concern and other-concern clash. Imagine someone strolling across campus and seeing a man menacing another person with a knife. Will he step in or keep his distance? In this circumstance, his desire to help the other person (other-concern) is at odds with his want to protect himself from the situation's danger (self-concern), and you must choose which aim to priorities.

There will be many more examples of the motives of self-concern and other-concern, both functioning together and against one other.

5.3. THE SOCIAL SITUATION CREATES POWERFUL SOCIAL INFLUENCE

When people are asked what they value the most, they frequently cite their social situation—that is, their interpersonal ties (Baumeister & Leary, 1995; Fiske & Haslam, 1996). People rely on others to help them get the job done when they collaborate on a class project, volunteer at a homeless shelter, or serve on a jury in a courtroom trial.

Humans form social relationships with those individuals and expect them to assist them in achieving their objectives. The value of others is evident in every element of human life—people teach one other what they should and shouldn't do, think, and even like and detest.

They are influenced by people who are not physically present but are nonetheless part of our thoughts and feelings, in addition to the people with whom they are actively interacting. Consider a person traveling home late at night on a remote country road.

There are no vehicles in sight in any direction, and the view extends for miles. They come to a stop sign. What will they do? They will most likely stop or slow down at the sign.

They do so because the conduct has become internalized: even when no one is watching, people are still influencing them—they've learned about society's rules and laws, what's proper and bad, and they tend to follow them. Every day, humans carry their own particular social situations with them—their interactions with their parents, instructors, leaders, authorities, and friends.

Although individuals' qualities contribute, social situations are generally a bigger determinant of behavior than personality, according to a key premise of social psychology that will be repeated throughout this book.

When social psychologists examine an event like the Holocaust, they are more likely to focus on the situation's qualities (e.g., the strong leader and the group pressure generated by other group members) rather than the perpetrators' characteristics. One will see, for example, that even ordinary individuals who are not terrible or evil in any way can be put in positions where an authority figure can induce them to engage in evil activities like applying potentially lethal doses of electrical shock (Milgram, 1974).

Social psychologists have established that often, one does not notice how essential the social setting is in shaping behavior, in addition to discovering the surprising amount to which human behavior is influenced by the social situation. One frequently mistakenly believes that people act completely independently of outside forces. It's easy to think that those who do extreme crimes, such as terrorists or suicide cult members, are uncommon or excessive. Despite this, numerous studies suggests that these actions are caused more by the social setting than by individual features, and that focusing solely on individual attributes is incorrect (Gilbert & Malone, 1995).

Perhaps no stronger illustration of the powerful influence of social circumstances can be found in research demonstrating the enormous importance that others play in our physical and mental health. *f*C (Diener, Suh, Lucas, & Smith, 1999; Diener, Tamir, & Scollon, 2006).

5.4. SOCIAL INFLUENCE CREATES SOCIAL NORMS

In certain circumstances, social influence occurs without the clear purpose of one person to affect another, such as when humans learn about and adopt the views and habits of others around them, frequently without realizing it.

When a young child inherits his or her parents' opinions and values, or when someone begins to adore jazz music without realizing it because a roommate plays it frequently, social influence occurs.

In other cases, social influence is anything but subtle; it involves one or more individuals actively attempting to persuade others to change their beliefs or behaviors, as evidenced by jury members attempting to persuade a dissenting member to change his or her mind, the use of a popular sports figure to persuade children to buy certain products, or cult leaders sending messages to their followers encouraging them to engage in the behaviors required of the grotesque.

The establishment of social norms—ways of thinking, feeling, or behaving that are shared by group members and deemed proper by them—is one result of social influence (Asch, 1955; Cialdini, 1993). Customs, traditions, standards, and rules, as well as the group's overall ideals, are all examples of norms. Humans learn what others actually do ("people in the United States are more likely to eat scrambled eggs in the morning and spaghetti in the evening, rather than vice versa"), as well as what we should and shouldn't do ("do unto others as you would have them do unto you"). Standards exist for practically every social conduct, and these norms have a significant impact on our actions.

5.4.1. Different Cultures Have Different Norms

Human culture contributes significantly to the social standards that guide our daily actions and establish social impact. A culture is a collection of people who live in a specific geographical area and share a set of social standards, such as religious and family values and moral convictions (Fiske, Kitayama, Markus, & Nisbett, 1998; Matsumoto, 2001).

Through instruction, imitation, and other forms of social transmission, the culture in which people live influences their thoughts, feelings, and behavior (Mesoudi, 2009). It is not an exaggeration to argue that culture, like evolutionary experience, shapes human life.

Cultures differ in terms of the particular norms that they value and use to govern group members' conduct. According to social psychologists, there is a fundamental difference in social norms between Western and East Asian cultures (including the United States, Canada, Western Europe, Australia, and New Zealand) (including China, Japan, Taiwan, Korea, India, and Southeast Asia).

Western cultural norms are generally oriented toward individualism—cultural norms prevalent in Western civilizations that emphasize self-improvement and independence. Children in Western societies are encouraged to acquire and appreciate a sense of personal identity and to regard themselves as distinct from those around them.

Children in Western cultures like receiving gold stars on their projects and receiving the highest mark in the class (Markus, Mullally, & Kitayama, 1997).

Adults in Western cultures are often motivated to promote their own personal achievement in comparison to (or even at the detriment of) others. When asked to describe themselves, people from Western cultures typically say they desire to "do their own thing," prefer to live their lives independently, and base their pleasure and self-worth on their own personal accomplishments. In summary, Western societies place a premium on self-interest. East Asian cultural norms, on the other hand, are more concerned with other people. These standards suggest that people should be more intrinsically related to others and, as a result, are more inclined toward interdependence, or collectivism.

Children in East Asian cultures are encouraged to prioritize creating harmonious social interactions with others, with the dominating norms relating to group cohesion, connectedness, and family duty and responsibility.

When asked to characterize themselves, representatives of East Asian cultures say they are very concerned about the interests of others, including close friends and colleagues.

Shinobu Kitayama and his colleagues discovered that East Asians were more likely than Westerners to experience happiness as a result of their social ties, but Westerners were more likely to experience happiness as a result of their own personal successes.

Other studies have looked into other cultural distinctions, such as variances in time orientations. Some cultures are more concerned about arriving and departing on time, while others are more flexible with time (Levine & Norenzayan, 1999).

According to Levine and colleagues (1999), "the pace of life" was fastest in Western countries (but also in Japan) and slowest in economically developing countries, as measured by average walking speed in downtown locales and the speed with which postal clerks performed a simple request.

It has also been suggested that persons in different cultures are constrained differently by social norms and practices, rather than being free to express themselves without respect for social norms (Gelfand et al., 1996). There are also cultural differences in terms of personal space, such as how near people stand to one other when chatting, as well as communication patterns.

Cultures and cultural differences are important to be aware of, at least in part because people from various cultural backgrounds are increasingly coming into contact with one another as a result of increased travel and immigration, as well as the development of the Internet and other forms of communication.

There are many different ethnic groups in Canada, for example, and the share of the population from minority (non-White) groups is growing year after year. Over the next few decades, minorities will make up a considerably higher proportion of the total new workers in Canada. Foreign-born Canadians account for roughly 21% of the population, easily the highest among the

G8 countries. By 2031, visible minorities are expected to account for 63% of Toronto's population and 59% of Vancouver's (Statistics Canada, 2011). While these changes have the potential to improve cultural understanding and productive engagement, they may also result in unwelcome social strife.

Cultural differences are an important aspect of a basic understanding of social psychology, and we shall return to this issue frequently in this book.

5.5. ENVIRONMENTAL IMPACT OF ECONOMIC GROWTH

Economic growth means an increase in real output (real GDP). Therefore, we are likely to see costs imposed on the environment with increased output and consumption.

As a consequence of economic growth, non-renewable resources are consumed more, pollution levels are higher, global warming increases, and potential habitat loss is likely.

Nevertheless, not all forms of economic growth cause damage to the environment. Real incomes are rising, which means individuals have greater means to protect the environment and mitigate the harmful effects of pollution. Furthermore, economic growth can lead to higher productivity and less pollution when technology is improved.

5.5.1. External Costs of Economic Growth

- **Pollution:** Increased consumption of fossil fuels can lead to immediate problems such as poor air quality and soot, (London smog of the 1950s).
- The Clean Air Act–which limits coal burning in city centers–has mitigated some of the worst problems of burning fossil fuels demonstrating how economic growth can reduce pollution.
- **Less Visible and Diffuse Pollution:** Unlike smog, the effects of increased CO_2 emissions are less immediately obvious. Therefore, policymakers have less incentive to address it. Global warming and erratic weather are attributed to CO_2 emissions. All of these issues indicate long-term environmental costs are increasing with economic growth–not just for the present, but for future generations.

- **Damage to nature:** Air/land/water pollution causes health problems and can damage the productivity of land and seas.
- **Global Warming and Volatile Weather:** Climate change can lead to rising sea levels and volatile weather patterns, causing significant economic losses.
- **Soil Erosion:** Deforestation resulting from economic development damages soil and makes areas more prone to drought.
- **Loss of Biodiversity:** Economic growth leads to resource depletion and loss of biodiversity. The ecological system carrying capacity for the economy may be adversely affected by this. However, it is unclear how much of a cost this is since the benefit of lost genetic maps may never be known.
- **Long-term Toxins:** Economic growth creates long-term waste and toxins, which may have unknown consequences. Economic growth, for example, has led to increased use of plastics, which do not degrade when thrown away. Consequently, there is an ever-increasing stock of plastic in the oceans and environment, which is unsightly and harmful to wildlife.

5.6. OTHER MODELS OF A LINK BETWEEN ECONOMIC GROWTH AND THE ENVIRONMENT

5.6.1. Limits Theory

This proposes that economic growth will harm the climate, and harm will itself begin to go about as a brake on economic growth and will drive economies to manage economic damage. All in all, the climate will compel us to care for it. For instance, assuming that we run down normal assets, their cost will rise and this will make a motivation to track down other alternatives.

5.6.2. New Toxics

This is a more cynical theory, recommending that economic growth prompts an always expanding scope of harmful results and issues, a few issues might get tackled, however, they are offset by fresher and additional squeezing issues which are troublesome if impossible to overturn (Figure 5.3).

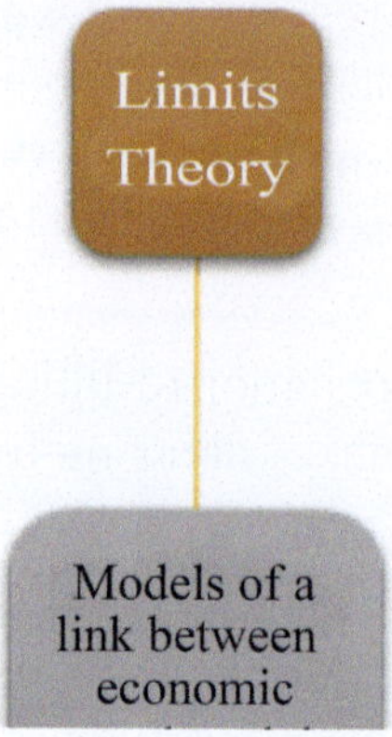

Figure 5.3. Models of a link between economic growth and the environment.

This model has no confidence that the unregulated economy will tackle the issue since there is no responsibility for quality and a significant number of the impacts are heaping onto people in the future; these future impacts can't be managed by the ongoing cost component.

5.6.3. Race to the Bottom

This suggests that in the early stages of economic growth, there is little worry about the climate and frequently nations sabotaged natural guidelines to acquire an upper hand–the motivator to complementary lift on others' endeavors. Be that as it may, as the climate progressively deteriorates, it will hesitantly drive economies to diminish the most exceedingly awful impacts of ecological harm. This will dial back natural corruption yet not converse with past patterns and trends.

Some ecologists argue economic growth invariably leads to environmental damage. However, some economists argue that economic growth can be consistent with a stable environment and even improvement in the environmental impact. This will involve:

- **A shift from Non-renewables to Renewables:** reports suggests that renewable energy is becoming cheaper than more damaging forms of energy production such as burning coal and in 2018–this has led to a 39% drop in new construction starts from 2017, and an 84% drop since 2015.
- **Social Cost Pricing:** If economic growth causes external costs, economists state it is socially efficient to include the external cost in the price (e.g., carbon tax). If the tax equals the full external

cost, it will lead to a socially efficient outcome and create a strong incentive to promote growth that minimizes external costs.

- **Treat the Environment as a Public Good:** Environmental policy which protects the environment, through regulations, government ownership and limits on external costs can, in theory, enable economic growth to be based on the protection of the environmental resource.
- **Technological Development:** It is possible to replace cars running on petrol with cars running on electricity from renewable sources. This enables an increase in output, but also a reduction in the environmental impact. There are numerous possible technological developments which can enable greater efficiency, lower costs and less environmental damage.
- **Include quality of Life and Environmental Indicators in Economic Statistics:** Rather than targeting GDP, environmental economists argue we should target a wider range of living standards + living standards + environmental indicators. (e.g., Genuine Progress Indicators GPI).

5.7. THE ECONOMIC IMPACTS OF ENVIRONMENTAL POLICY

The essential objective of the ecological approach is to guarantee that the indigenous habitat is overseen and utilized reasonably, and to keep away from the break of any basic edges past which abrupt, sensational, or irreversible changes might happen; or past which replacement by other regular resources or different variables of creation is absurd.

This gets the many advantages that society gets from the climate, as a contribution to financial activities and as a driver of individuals' prosperity and personal satisfaction by its own doing.

5.7.1. Investment and Innovation

Technological advancement and the improvement of new information are significant drivers of monetary development and are critical elements in guaranteeing that the shift to earth maintainable development occurs at a cost to the economy. New information is produced by R&D–which could be supported openly (when a portion of the advantages build to society overall) or secretly (where the advantages are to a great extent private and on which the specialists can create again). Government strategy, and explicitly natural approach, play a significant part in boosting mechanical advancement and development. Arrangements pointed toward guaranteeing that natural data sources are evaluated accurately are probably going to spike organizations improve to diminish costs.

For instance, Reid, and Miedzinski (2008) observe that administration strategy is a significant driver of green development. In particular, government strategy can support natural advancement through 'request pull' arrangements, for example, guidelines or public obtainment that increment interest for development, and 'supply-push' approaches, for example, endowments, and tax reductions for research.

What's more, a predictable and rational natural strategy can give more noteworthy sureness about the worth of ventures and boost ecological R&D further towards the socially ideal level. Doorman and van der Linde (1995) track down proof of this impact, as well as the impact of ecological guidelines in lessening latency and bringing issues to light among firms of shortcomings in their creative processes (Figure 5.4).

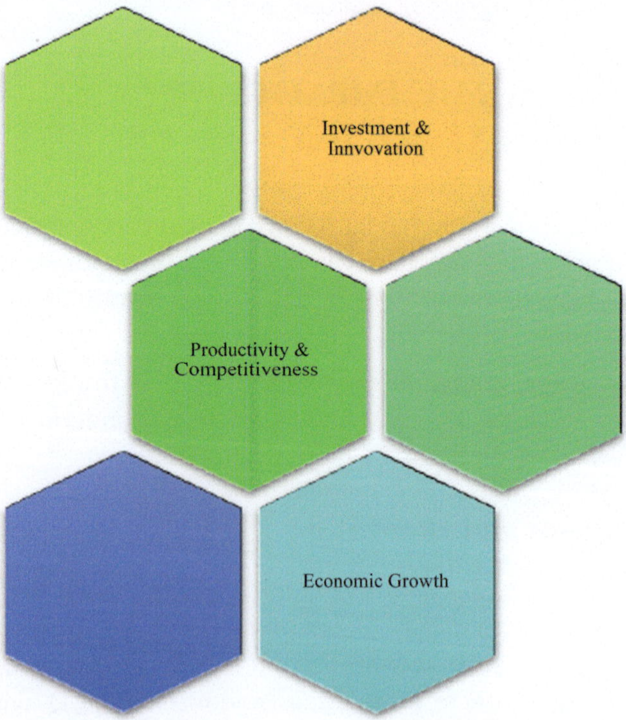

Figure 5.4. The economic impacts of environmental policy.

Requiring consistency with ecological guidelines has likewise been found to increment advancement–for instance, Jaffe, and Palmer (1997) view an expansion in consistence use as related to a general short-run expansion in R&D.

In any case, we want to survey the effect of natural strategy on speculation and development from the perspective of the entire economy. Will streams of interest in ecological development 'swarm out' existing speculations, and is there an open-door cost from putting resources into natural advancement?

Because of the wide scope of elements impacting the economy, it is troublesome to arrive at firm resolutions on the impact of natural arrangement on by and large degrees of R&D venture.

For instance, early investigations of the effect of environment strategies found that they invigorated development in elective energy ventures, however, beat R&D in non-energy areas down–prompting a constriction in all-out creation and diminishing the general pace of specialized progress.

In any case, later examinations have arrived at various resolutions. For instance, worldwide mechanical overflows might bring about a nation putting less in ecological R&D (as it benefits from R&D led abroad–keeping away from duplication and empowering venture somewhere else), bringing about an in all cases expansion in the degree of development.

Gerlagh (2008) finds that, as the number of input rises, information aggregation shifts from energy creation to energy-saving advancements, and recommends a rising level of innovative change per unit of the venture. Carraro et al (2009) and Carraro et al (2009a) dissect environment strategy and find that interest in energy-related R&D don't prompt jamming out of interest in different areas, nor do they lead to a decay in degrees of human resources.

5.7.2. Productivity and Competitiveness

The literature on environmental regulation and productivity has identified several ways in which environmental regulation could lead to reductions in productivity.

For example:

- Benefits from environmental regulation may be non-market, and therefore not reflected in conventional measures of productivity.
- new technologies and production processes needed to comply

with the regulation may be less efficient in their use of market inputs than before;

- environmental regulation may lead to firms with higher levels of measured productivity being crowded-out; and
- it may be a disincentive for firms to invest in new technologies if they expect to be subject to more regulation in the future.

Empirical evidence, on the contrary, has found ecological guidelines to have a minor unfavorable effect, if any, on efficiency. A significant number of the investigations assessing the long run long-run of ecological guidelines on efficiency must produce results that are not measurably hearty and are profoundly reliant to show determinations.

The writing additionally proposes that the impact of the ecological guideline on efficiency relies upon the business viability, as well as the strategy for demonstrating productivity. Despite this, no review has tracked down enormous adverse consequences of ecological guidelines on general efficiency, either in the short or over the long haul.

As far as serious impacts of ecological strategy, Stewart (1993) observes that the degree of the guideline is certainly not a significant determinant of the intensity of a country, with no apparent change in global exchange design reaction to natural guidelines.

In any case, certain areas might be more defenseless than others, contingent upon the force with which they depend on the managed natural resource.

The negative effects of environmental regulation on productivity and competitiveness could be limited through:

- Early signaling and clear regulatory frameworks to allow firms to anticipate new regulations and take them into account in their business planning;
- Policies that stimulate R&D, such as subsidies and grants, that facilitate the introduction of new and improved technologies;
- Fiscally neutral policies, using any revenue raised to reduce other distortionary taxes in the economy;
- International agreements that reduce competitiveness impacts.

5.7.3. Economic Growth

Environmental policy can result in savings and benefits to businesses and industries–reductions in resource costs from improved resource efficiency; growth in expanding environmental industries and increased international competitiveness for 'first movers'; and lower risks to growth from improved business resilience to environmental shocks. It has also been noted that environmental policies need not have negative implications for investment.

What are the implications of environmental policy for economic growth?

Porter (1991) alludes to the possible struggle between development and the climate as a 'bogus polarity,' finding that a very much planned natural strategy can expand R&D into asset effective items and cycles, bringing about superior business seriousness and benefit. This hypothesis is known as the "Porter Hypothesis."

There is experimental proof to help the speculation. For instance, Meyer (1992, 1993) observe that there was no genuinely critical adverse consequence on development in US states with additional tough ecological guidelines. Albrecht (1998) tracks down proof to help the speculation while examining the impact of the Montreal Protocol on Danish and US firms.

Be that as it may, later exploration has tracked down blended results. Specifically, Frohwein, and Hansjürgens (2005) examine the Porter speculation with regards to the late EU REACH guideline. They view that as albeit a few firms might profit from the more severe guidelines, others would be more awful off.

Jorgenson and Wilcoxen (1990) track down that capital speculation to conform to additional severe natural guidelines eased back the US economy by 0.2% yearly somewhere in the range of 1974 and 1985 contrasted with the same old thing. Besides, the review presumes that the development impacts would have been much humbler had more powerful strategy decisions been made. In breaking down the monetary effect of the EU Emissions Trading Scheme, Peterson (2003) likewise tracks down a negative macroeconomic impact, yet impacts were viewed as similarly while possibly not more immaterial.

The Porter Hypothesis has its Critics. It has Been Argued that Porter Implicitly:

Assumes that firms willingly give up opportunities to improve not just their environmental performance but also their competitiveness; and does not take account of the fact that the scale of the growth benefits accruing from environmental policy also depends on developing regulation that is stringent but also efficient.

It likewise doesn't consider that a large number of the advantages of natural strategy are non-market–that is, they are not reflected in market costs. For instance, upgrades in degrees of biodiversity or the assurance of 'squander sink' administrations given by the seas and soils.

While these advantages might be esteemed by society, they are not completely reflected in proportions of monetary action and won't naturally convert into expansions in benefits and GDP.

While the natural approach can deliver positive development results–for instance, by boosting expanded R&D into asset proficient items and cycles that further develop in general business seriousness and productivity–it should be adjusted against a portion of different impacts of ecological strategy.

To be sure, where the ecological approach brings about an expansion in costs (for instance, water quality guidelines prompting higher sewerage rates for organizations) there may, in the short-run, be expenses for natural arrangement.

Accordingly, there could be a component of compromise between financial development in the close term and ecological approaches that safeguard regular resources, and monetary development, in the long haul. All the more, by, and large, approaches pointed toward valuing ecological assets accurately into monetary choices could bring costs up in the close term, and this would be considered against the development and more prominent proficiency in asset utilize that these arrangements can boost.

In any case, any close term expenses of executing ecological strategy should be seen with regards to the expenses of not making a move, or at least, contrasted with what might happen under a 'the same old thing' situation. In the long haul, the expense for the development of acting now to guarantee practical and proficient utilization of regular resources is probably going to be more modest than the expenses of inaction.

A large portion of the examination in such a manner has been directed at meeting environmental change targets. For instance, Stern (2006) finds that the expenses of lessening fossil fuel by-products to keep temperature climbing to inside 2°C would be around 1% of worldwide GDP in 2050. He analyzes this to the assessed cost of inaction of 5–20% of GDP. These discoveries have been built up by other later examinations. Comparative outcomes have been found for other ecological difficulties–for instance, that's what a new report gauges, that by 2050, combined misfortunes of environments administrations from biodiversity could be identical to around 7% of worldwide utilization.

In any case, the possible compromise between wanted natural results and monetary development in the close term features the significance of very much planned ecological approaches that tackle ecological issues in and proficient way and limit administrative weights on organizations and the economy.

5.8. SUSTAINING ECONOMIC GROWTH IN THE LONG-TERM

5.8.1. The Case for Economic Growth

Not with standing transient slumps and mishaps, the drawn-out pattern in monetary results throughout recent years has been unambiguously vertically. Ground-breaking thoughts and their transmission, in blend with the collection of work and capital, have empowered and supported monetary development.

Normal pay has significantly increased throughout recent years (with outright expansions in family pay for even the most minimal pay gatherings) and individuals are more taught, better, and have a better quality of living than at any other time.

The advantages of monetary development have not been limited to the UK or other high-level economies. As worldwide GDP has duplicated a 21-crease throughout recent years, it has worked on personal satisfaction and hauls incalculable millions out of neediness.

Monetary development stays crucial to help proceeded with enhancements in factors that influence individuals' prosperity, from wellbeing and work to schooling and personal satisfaction, and to assist the public authority with following through on a scope of strategy targets–financial, social, and ecological.

It is crucial for supporting proceeded with enhancements in material expectations for everyday comforts, for instance, by setting out business open doors and by establishing an appealing climate for a private venture.

Through the assessment framework, monetary development likewise upholds different variables influencing prosperity, for instance, through proceeding with enhancements in the arrangement of public administrations and in help for lower pay families that diminish destitution, further develop wellbeing results, and lead to more prominent instructive achievement.

The shift to low carbon and more asset productive economy will require essential changes in the construction of the economy. It will require an interest in new advancements and development, and speculations to supplant maturing foundations and decrease future dangers from ecological change.

Monetary development permits these requests to be met without fundamentally diminishing interests in different regions that make a difference in the prosperity of people and society. Looking past the UK, development furnishes creating economies with the potential chance to work on the personal satisfaction of their residents, creating organizations and businesses, raising earnings and giving how they can address the natural difficulties they face.

Through the exchange, speculation, help, and settlement streams, proceeded development in cutting edge economies plays a significant part to play in diminishing neediness and increasing expectations of living across the world.

5.8.2. Natural Capital and Sustainable Economic Growth

The process of wealth creation is generally described in terms of combining factors of production to produce goods and services. Some of these goods and services are consumed, while others are used to enhance the capital stock.

However, this formulation of output does not fully account for the role of natural capital in the production process and provides an incomplete picture of the contribution of natural capital to economic growth and wealth creation.

The key factors of production to be considered in the context of economic growth are:

- Produced capital – usually man-made capital such as machinery and infrastructure;
- Human capital – such as labor effort, skills, education, and experience;
- Natural capital – the raw materials and services provided by the

natural environment, such as wood, minerals, water, nutrient recycling; and

- Social capital – whilst definitions of social capital differ, it generally includes institutions and ties within communities.

An expansion in the amount of these elements of creation increments monetary result; for instance, through an expansion in the workforce or through the advancement of hardware and constructed framework.

Likewise, mechanical advancement and upgrades like these elements of creation further develop efficiency and increment yield; for instance, mechanical advancement and the gathering and use of information permit new and better approaches to consolidate the different variables of creation to deliver yield.

5.9. CONCLUSION

The chapter explores the vast differences in eastern and western societal norms, the dependence of behavior on one's social setting is stressed upon, and the importance of social influence on behavior. In conclusion, contrary to how one might think human behavior is not just dependent on the individual, it is influenced by multiple factors such as a person's family, friends, upbringing, social setting and even nationality.

REFERENCES

1. Breed, M., & Moore, J., (2022). Comparative social behavior. *Animal Behavior*, 489–530 [online] Available at: https://www.sciencedirect.com/science/article/pii/B9780128195581000142?via%3Dihub (accessed on 27 June 2022).
2. Georgoulas, N., (2021). Social-behavioral development. *7th International e-Conference on Studies in Humanities and Social Sciences: Conference Proceedings* [online] Available at: https://www.sciencedirect.com/science/article/pii/B9780128195581000142?via%3Dihub (accessed on 27 June 2022).
3. Petrakis, P., Kafka, K., Kostis, P., & Valsamis, D., (2021). Social behaviors. *Greek Culture After the Financial Crisis and the COVID-19 Crisis*, 127–137 [online] Available at: https://link.springer.com/chapter/10.1007/978-3-030-81018-4_8 (accessed on 27 June 2022).
4. Stangor, D., Jhangiani, D., & Tarry, D., (2022). *Defining Social Psychology: History and Principles*. [online] Opentextbc.ca. Available at: https://opentextbc.ca/socialpsychology/chapter/defining-social-psychology-history-and-principles/ (accessed on 27 June 2022).
5. Vijayalakshmi, N., (2019). Behavior modification techniques–an awareness study. *SHANLAX International Journal of Education, 7*(2), 20–24. [online] Available at: https://files.eric.ed.gov/fulltext/EJ1245283.pdf (accessed on 27 June 2022).

CHAPTER 6

Sociobiology–Behavioral Ecology

CONTENTS

The comprehensive study of how and why behavioral mechanisms and processes govern animals' interactions with their biotic and abiotic environments, consequently shaping numerous ecological and evolutionary processes, is known as behavioral ecology.

The behavior of animals connects physiological and molecular genetic processes with the external environment of organisms, putting behavioral ecology at the heart of how evolution works in nature. Behavioral ecology evolved from two academic disciplines: ethology and sociobiology.

6.1. INTRODUCTION

Ethology was the first organized effort to study behavior scientifically, although it was mostly descriptive and field-oriented. Scientists' replies to Niko Tinbergen's structure of animal behavior around four distinct topics: cause, function, development, and evolutionary history, changed this discipline (or phylogeny).

Sociobiology, although being a politically sensitive issue for several decades, introduced clear cost-benefit assessments and a functional evolutionary viewpoint on how animals interact in natural surroundings.

Although this field was short-lived, possibly due to its contentious formulation of modern human behavior and how some people interpreted it in terms of cultural norms, sociobiology still aided in the emergence of behavioral ecology as a core evolutionary discipline that viewed organisms as an integral part of natural habitats.

Because behavioral ecology is a fundamentally evolutionary field, most of what practitioners investigate revolves around evolutionary causes and adaptive explanations of behavior. Behavioral ecologists have made significant contributions to theoretical explanations of evolutionary mechanisms, as well as to demonstrating how evolution operates on modern time scales.

It is critical to understand that behavioral ecologists aim to explain the functional evolution of behaviors in all sorts of species, not only animals, resulting in a taxonomically vast area. There are several approaches to deciphering the behavioral mechanisms and processes of ecology and evolution. These are classified as:

- Foraging and survival tactics;
- Sociality and social structure;
- Breeding and reproduction;

- Mobility and dispersal; and
- Communication and cognition.

The word "sociobiology" is also used to refer to contemporary evolutionary approaches to human behavior that maintain the behavioral focus to which Wilson alludes. While influenced by earlier sociobiology, most modern projects studying the evolution of behavior appear to have developed a variety of theoretical and methodological features of their own. Perhaps the best way to understand sociobiology, in this case, is as a "phylogenetic" ancestor of these projects.

Human behavioral ecology (HBE), evolutionary psychology, and, more recently, dual inheritance theory are all "descendants" of sociobiology. These efforts are now referred regarded together as the "evolutionary social sciences."

Different theoretical perspectives and methodology choices have become entrenched in each project over time, increasing disparity; divisions also partly fall along disciplinary lines, with human behavioral ecologists tending to be anthropologists, evolutionary psychologists tending to be cognitive psychologists, and individuals working in dual inheritance theory tending to have some background in population biology modeling.

However, all three evolutionary social sciences maintain significant aspects of the project, most notably its methodologically adaptationist approach. Because of its emphasis on the evolution of behavior, HBE is the most commonly referred to as "sociobiology" of the three (evolutionary psychology focuses on the evolution of psychological mechanisms or modules such as those for cheater detection; dual inheritance theory studies the cultural evolution of traits that are socially learned)

On the three primary definitions of "sociobiology." First, it will quickly introduce the relatively uncontroversial non-human animal behavioral ecology known as "sociobiology" by Wilson and others.

Second, it will discuss and assess the characterization of "Pop Sociobiology," which was attacked by a number of scientists and philosophers in the mid to late 1970s. Third, it will discuss the major characteristics of contemporary HBE. The last section will cover some of the most common critiques leveled towards sociobiology and behavioral ecology in the literature, as well as possible solutions.

What animals do and why they do it is referred to as behavior. Behavioral ecology studies the development of animal behaviors that allow them to adapt to and prosper in their environments. There are two types of behavior: learnt and instinctual. Instinctive behavior is a hereditary pattern that is passed down from generation to generation.

A spider, for example, does not need to witness another spider spin a web to know how, where, and when to do so. This knowledge is natural in the spider, allowing it to carry out many of its life activities without having to think about them. The drawback of instinct is that it is rigid and does not allow the animal to adjust its behavior when it is no longer suitable.

When attacked, the armadillo's innate upward leap functioned perfectly until it discovered a new environmental hazard—the vehicle. Learned behavior, on the other hand, is the consequence of collected and absorbed experience throughout a lifetime that allows the animal to adjust to unanticipated changes.

A behavioral ecologist investigates behavioral patterns that are midway between instinctual and taught. They are as follows:

- **Reflex:** It is a quick automatic reaction to stimuli. When threatened, hedgehogs reflexively curl into a ball.
- **Conditioned reflex:** An instinctual response that may be conditioned to occur in a variety of situations. When a whip is flicked at a racehorse, it links the whip with its traditional predator, a huge cat clawing at its back.
- **Migration: A** Seasonal relocation to a more suitable summer or winter climate. The monarch butterfly's migration, which spans hundreds of miles and two generations, is one of the most incredible. The children are genetically predisposed to return to the farms that their parents abandoned.
- **Hibernation and estivation:** Some animals adopt a condition of torpor, or a reduced metabolic rate similar to sleep, in order to endure harsh winters or scorching, dry summers.
- **Imprinting:** A baby animal's memory of the form, sound, or scent of its parents or birthplace within the first few hours after birth. If the parent is not around, the infant will imprint on the first item it sees, leading to the sight of ducklings who believe people are their parents or kittens that have imprinted on dogs.
- **Courtship:** The complex signals and rituals that allow male-female connections to form for mating purposes. These actions ensure the intent and, as a result, the safety of both mates, who

may fight or consume an approaching mate if these behaviors are not seen.

- **Mimicry**: The transformation of a harmless animal into a harmful animal's appearance or behavior. The viceroy butterfly resembles the toxic monarch butterfly, which most birds are genetically designed to avoid.
- **Preadaptation**: is the combination of innate and learnt behavior. Purple martins, which used to nest on cliffs, have learnt to exploit man-made buildings to expand their habitats.

Behavioral ecologists who study animals in their natural habitats describe countless instances of seeing them meet a new scenario and devise a new reaction. E. O. Wilson, a Harvard scientist, remembers observing numerous beavers whose dam had been damaged come up with a solution to the problem.

Because the water flow was too powerful for their instinctual skills to stop, the beavers devised a novel and successful method of mending the dam with sticky underwater mud and debris. Wilson believes that this demonstrated the beavers' capacity to appraise an issue and solve it via reasoning.

Behavioral ecology investigates how adaptations emerge from selecting pressures in various ecological contexts, including novel selective environments created by animal behavior. Behavioral ecology investigates the evolutionary causes and ecological implications of behavior at the crossroads of ecology, evolution, neurology, and genetics (Figure 6.1).

Figure 6.1. Gibbons on branches and vines.

Source: image by Wikimedia Commons

6.2. SOCIOBIOLOGY AS BEHAVIORAL ECOLOGY

The initial meaning of "sociobiology" is a name for a body of study that is now more often known as behavioral ecology. Behavioral ecology is a field that attempts to understand animal behavior by applying evolutionary theory, particularly adaptationist approaches. Take, for instance, the female parasitic jewel wasp.

This wasp lives and lays its eggs in blowfly pupae; when a female wasp of this species matures, she mates with one of the males who is also inside her pupa, and then emerges and flies off to lay her eggs in another pupa. The jewel wasp is haplodiploid, and females may choose the sex of their progeny by deciding whether or not to fertilize the egg when the female lays it.

What's fascinating about this wasp is that the ratio of females to males she lays depends on whether she arrived at the pupa first. If she is the first to arrive at the pupa, she deposits a large number of eggs, predominantly females (approximately 91.3%) and some males (about 8.7%).

However, if she lays eggs in a pupa after another female has been there (a process known as super parasitism), she produces many more males than daughters. Behavioral ecologists want to know why animal behaviors like these have such distinctive characteristics.

Behavioral ecology is frequently defined by its ties to older "ethology," the most notable practitioners of which are arguably Konrad Lorenz and Niko Tinbergen. Tinbergen notably identified four sorts of questions that an ethologist may ask in order to comprehend a pattern of animal behavior, such as the jewel wasp's egg-laying approach.

First, the ethologist may inquire about the immediate origins of the behavior, such as what neurological or psychological systems drive the wasp to deposit her eggs in the manner she does. What mechanisms are accountable for the mechanism's development? What developmental mechanisms produce the neuronal tissue that is directly responsible for egg laying behavior?

Second, the ethologist could ask two ultimate evolutionary questions about the behavior: what is the evolutionary function of the behavior, i.e., how does laying eggs in that pattern contribute to the wasp's reproductive success, the reproductive success of its ancestors, and thus to the behavior's history of evolution by natural selection in the jewel wasp population (if that is its history)? Also, what is the phylogeny of the behavior, i.e., where did it emerge in the jewel wasp's evolutionary tree?

Apart from ethologists, behavioral ecologists studying animal behavior initially decided to concentrate on the third question, evolutionary function, at the expense of the others. However, such a focus by behavioral ecologists on questions of function at the expense of questions of causation, development, and phylogeny (especially development) was highly problematic, because the four questions cannot be adequately answered independently.

Development limits the evolution of biological qualities in a variety of ways, particularly behavior because behavior is dependent on development at both the neurological and psychological levels. Ecologists of behavior, Nevertheless, for quite a long time, researchers avoided such investigations as part of their study of the evolution of behavior due to the complexity of the developmental resources contributing to the behavior.

According to Grafen (1984), the problem with developmental approaches to behavior is that they are difficult and time-consuming to carry out, especially if the only or primary purpose of the developmental work is to ensure that the trait in question has an evolutionary history that can be understood using adaptationist methods.

Grafen contends that if behavioral ecologists had to conduct developmental or psychological studies in order to be reasonably confident in their conclusions about the evolutionary history of traits, they would likely never progress to evolutionary studies because such developmental and genetic work would take a very long time.

Worse, such investigations may have no intrinsic interest and so constitute a waste of the scientist's time and resources—not all features have an interesting or instructive developmental history or are created by fascinating and unexplored psychological mechanisms. Instead, Grafen contends that behavioral ecologists should employ what he refers to as the "phenotypic gambit" when investigating the evolution of behavioral features.

This entails assuming that the developmental, psychological, and genetic proximate causes of a behavioral trait, such as the jewel wasp's egg-laying strategy, are of the type that will allow scientists to "pretend" that they are very straightforward when building evolutionary models, and still end up with models that give precisely the right responses about the fundamental steps, conditions, and constraints trying to act on that characteristic throughout its evolution.

However, since Grafen's article, there has been a shift in non-HBE away from the phenotypic gambit and toward paying attention to proximate mechanisms, not only because these enlighten whether a behavioral strategy could evolve by natural selection in any particular case, but also because the nature of the fitness costs and benefits that accrue to an organism sometimes depends on how the adaptive problem they face is solved.

For instance, Lotem et al. (1995) cite the sort of learning method utilized by parents of great reed warblers to identify the existence of cuckoo eggs in their nests, which restricts the reliability of their detection of cuckoo eggs. Similarly, Shuker, and West (2004) argue that the constraints on the parasitic jewel wasp's egg-laying strategy cannot be properly understood unless the limits on the detection mechanisms the wasp uses to determine the presence or absence of other females versus other females' eggs in each area are considered.

The optimality model is perhaps the most important analytical tool of the behavioral ecologist; however, frequency-dependent models that do not assume population fixation of a trait, such as evolutionary stable strategy, and dynamic models, such as those used in studying life history traits, are also used.

Optimality models are intended to demonstrate which of a set of plausible alternative behavioral methods would maximize a local fitness currency given a variety of ecological variables and limitations implicitly or explicitly stated in the model's mathematical framework.

These conditions are supposed to represent the trait's selection conditions (the conditions that drive its evolution by natural selection–Brandon, 1990); however, they are usually derived by observing the conditions in which the organism is currently living, because these are assumed to have remained largely unchanged from the conditions in which the organism evolved. Werren (1980) proposed precisely such a model for the jewel wasp above; the model argues that the shift in the ratio of sons to daughters when a female wasp finds an occupied pupa is due to the possibility of mating with the first wasp's daughters.

If just the first wasp lays her eggs on the pupa, the hatching daughters mate with their brothers, and all of their progeny are completely descended from the original wasp. This means that the first wasp must produce relatively few sons—only enough to mate with all of the females; the first wasp's second-generation reproductive success is optimized by increasing the number of daughters.

However, when there is a super parasitic wasp, the super parasitic wasp's sons can mate with the first wasp's vast number of daughters. The super parasitic wasp must increase the number of females in the pupa that mate with her sons rather than their brothers while minimizing competition among her boys.

If she just has a few eggs to lay, it makes sense for her to lay nearly all males; as brood sizes increase, it makes sense to progressively increase the number of daughters in the mix. Warren estimated the ideal male and female percentages for each brood size. His findings of male to female ratios follow the model's predictions very well, but not completely.

Various philosophers and behavioral ecologists disagree over what optimality models such as Werren's are used to illustrate. According to Orzack and Sober (1994), behavioral ecologists such as Werren employ optimality models to demonstrate that the qualities in issue are adaptations, i.e., that the traits developed by natural selection.

Furthermore, Orzack, and Sober consider optimality models to be censored models, that is, they are designed to determine the outcome of natural selection in a population where natural selection is the only important force, and thus demonstrate, if indeed the model accurately anticipates the similar phenomenon of the trait under study, that natural selection has been the most influential factor within the evolution.

The theory is that Werren's model's relative success demonstrates that natural selection was the most significant factor in the evolution of the jewel wasp's egg-laying strategy—other forces may have been involved, but they are not as causally important.

However, various other philosophers have argued that such models do not make such strongly adaptationist assumptions; Potochnik (2009), for example, argues that optimality models are only designed to show the general role that natural selection and other forces (such as developmental constraints) are playing, and are not designed to test strong adaptationist hypotheses about the behavioral strategies in question—i.e. in our case above, the success of the model is intended to show that natural selection was acting on the jewel wasp's strategy, and that the way it was acting on the jewel wasp's strategy was via the effects of competition between the male wasps in the pupa, the variations in brood size and the other conditions and constraints to which Werren's model appeals.

The model's success does not imply that natural selection was the sole mechanism at work in the evolution of the jewel wasp's approach. Other objections to the censored model view may be found here. Furthermore, some behavioral ecologists have asserted that they do not utilize optimality models to assess whether or not natural selection is acting on a trait, but rather presume natural selection is operating in order to test hypotheses regarding the circumstances and constraints acting on the characteristic. This might be problematic if it implies that behavioral ecologists accept the assumption that natural selection optimizes most features without ever fully testing it.

However, one reading of Parker and Maynard Smith (1990) is that behavioral ecologists do not have an unshakeable belief that natural selection is an optimizer. Instead, they argue that evaluating ideas regarding the functioning of natural selection is not what behavioral ecologists are interested in.

In order to test other hypotheses about a trait, behavioral ecologists "pretend as if" it is an adaptation—they "pretend as if" the trait is an adaptation, and use that assumption to detect deviations from optimality that might reveal interesting tradeoffs and constraints—the nature of the deviation can suggest what might be missing from the model.

As a result, when the models first fail, scientists do not hunt for alternative non-adaptationist theories, but rather for new restrictions, tradeoffs, and situations under which selection may be working. Giraldeau (1997), for example, reveals departures from optimality in the formation of photopigments in rods (rather than cones) in the retinas of fish that reside in certain environments types of marine habitats in which a change in photosensitivity would be beneficial.

Giraldeau interprets this anomaly as the presence of a molecular limitation on rod cell development. Repeated failures to fill in the gaps in these optimality models may eventually lead to the scientists involved investigating alternate, non-adaptationist interpretations.

To undertake this work, behavioral ecologists do not need to be dedicated to the strong belief that natural selection is always the most significant factor in the formation of features; this heuristic simply requires the belief that natural selection generally locally optimizes (i.e., optimizes only relative to possibly severe constraints).

Therefore, another approach to explain how behavioral ecologists use optimality models is that they are interested in providing a functional description of the features they are examining (for a similar view in terms of choice patterns rather than behavioral methods).

One of the concerns that a behavioral ecologist may have been how to precisely explain the nature of the behavior under consideration: what component behavioral dispositions (from among all of the organism's behavior) comprise each strategy, and how those dispositions hang together as an adaptation.

They must describe the link between the organism's behavior as well as its environment, as well as how this adds to the organism's fitness.

Wasp in the jewel the behavioral ecologists have observed varying egg sex ratios in wasps on different occasions. The question is how to make sense of these various sex-ratio laying behaviors: the guess is that these various behaviors are linked together in some way, as manifestations of a specific egg-sex-ratio adjusting strategy; and that having this specific strategy, this specific pattern of egg-sex-ratio laying behavior benefits the female wasp as a whole.

But why should a behavioral ecologist believe that any specific functional description is correct?

A behavioral ecologist can use an optimality model to find and evaluate such a functional description in two ways. To begin, an optimality model (explicitly or implicitly) comprises a strategy set, which is a collection of descriptions of potential behavioral strategies; the model selects one of these strategy descriptions as the most adaptable.

According to Werren's model, the strategy set would be a (implicit) collection of descriptions of the many sex-ratio modifying techniques that a jewel wasp may employ (i.e., each strategy involves a different set of choices about sex ratios in a full range of relevant conditions like parasitism, super parasitism and different brood sizes).

The notion is that an optimality model can determine which member of this collection is best. Werren's approach is meant to choose the ideal sex ratio egg laying strategy description from the strategy set based on this interpretation of what behavioral ecologists are doing. The hypothesis is that this description of the most adaptable approach is valid.

If this is the proper description, Werren may now estimate what son/daughter ratios he should see the wasp lay in previously unseen circumstances. These predictions allow him to verify whether the description is valid—if the predictions are correct, Werren has proven that the description of the specific sex-ratio adjustment method obtained from his model is the proper description of how the wasp is behaving.

Second, the model's explanation of the strategic plan is not purely descriptive of the strategy as a series of behavioral temperaments (i.e., dispositions to lay different sex ratios under different conditions), but of the relationships between those dispositions and multiple environmental characteristics which decide the health and wellbeing of those dispositions.

In the case of the jewel wasp, for example, the presence of other female competitors, flexibility in brood size, the ability of sons to mate multiple times, the impossibility of son migration between pupae, etc., are environmental features that determine whether any potential jewel wasp egg laying strategy is maximally highly adaptable, and a very well confirmed optimality model could really demonstrate how.

Accordingly, depending on whether or not the model is valid, the description of the strategy provided by a successful example is (implicitly or explicitly) a functional explanation of the strategic approach representation of how much the behavior adds to the organism's fitness. For a more in-depth look at the importance of optimization modeling in biology.

6.3. POP SOCIOBIOLOGY

The second way that "sociobiology" has come to be understood is as a particular approach to understanding specifically human behavior which Philip Kitcher (1985) calls "Pop Sociobiology" (as opposed to his description of "narrow sociobiology" which is roughly equivalent to "behavioral ecology" above).

Pop Sociobiology is so-called because it is a view about how to study human behavior described in a variety of literature written by Wilson and others for a general, rather than an academic audience. In this literature, Wilson, and the other "Pop sociobiologists" present some speculative and preliminary sketches of how an evolutionary science of human behavior might proceed:

Wilson's main focus in On Human Nature (Wilson, 1978) and to a lesser extent the chapter of Sociobiology (Wilson, 1975) is to show that such a science is possible, to describe some of the techniques that might be used in pursuing it, and to sketch some possible evolutionary analyzes for certain particular human behaviors.

Because of its presentation in the popular press, "Pop Sociobiology" was probably important in shaping popular perceptions of the nature of sociobiology in general (for example, the Time article, "Why you do what you do" 1977, 110 (Aug 1), 54) and consequently drew the ire of the critics. Unfortunately, the intensity of this debate may have led to a certain amount of mischaracterization of the sociobiologists' views. This section will address the main concerns that the critics raised about Wilson's early "Pop" sociobiology, and discuss whether and how far these are fair descriptions of his views (Figure 6.2).

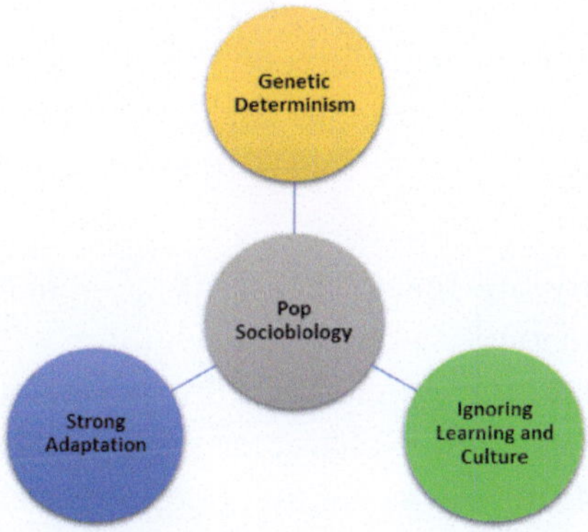

Figure 6.2. Pop sociobiology.

6.3.1. Genetic Determinism

Major critiques of sociobiology, such as Stephen J. Gould and the so-called "Sociobiology Study Group" (hereafter SSG), claim in a variety of articles that sociobiologists are strong genetic determinists. Wilson, for example, argues that there are specific genes "for" behavioral qualities such as indoctrinability, territoriality, warfare, and reciprocal altruism and that these genes are amenable to natural selection in a reasonably uncomplicated manner.

Indeed, the SSG argue that claiming that traits have a selective origin necessitates the existence of genes "for" them; Wilson's evident acknowledgement that traits may frequently have such a strong cultural component is said to be an error, because if this is true, then there must be genes "for" them. Since if that's really correct, evolutionary theory teaches us nothing about the genesis of such characteristics.

Similarly, Gould contends that sociobiologists are unaware that genes only develop features with the help of the environment.

6.3.2. Ignoring Learning and Culture

As a corollary to the charge that Pop Sociobiology is committed to genetic determinism, its central figures are frequently accused of being insensitive to the issue of learning and culture, i.e., to the fact that many traits in which they are interested are simply not subject to natural selection at all, and that this state of affairs may be common in humans.

However, Wilson, for example, clearly recognized the importance of culture in many behavioral traits; in fact, he believed that even minor genetic differences that made a difference in behavior could be exaggerated by acquired culture—this is known as the "multiplier effect" (although the multiplier effect's validity is seriously questioned–Maynard Smith and Warren, 1982).

Moreover, partially in reaction to his opponents' worries, Wilson went on to co-author Genes, Minds, and Culture with Charles Lumsden, an attempt to investigate the impact of cultural transmission on the nature and dissemination of behavioral characteristics, as well as the relationship between genes and culture. However, the book was heavily criticized.

The major problem was that the models offered in the book had little substance—the most fascinating aspects merely resulted from the assumptions incorporated into them, namely assumptions regarding the degree to which genes held culture "on a leash."

6.3.3. Strong Adaptationism

The third issue with Pop Sociobiology was its dependence on an extremely strong kind of adaptationism. Critics of sociobiology contend that sociobiologists are committed to a "Panglossian" adaptationism in both articles by the SSG (1976) and by Gould and Lewontin in their famous "Spandrels of San Marco" paper (Gould and Lewontin, 1979). While the "Spandrels" study is aimed at "adaptationists" in general, sociobiologists were specifically mentioned (for example, David Barash's (1976) work on jealousy in male bluebirds).

The focal allegations of the "Spandrels" paper were as per the following: that adaptationists treat all characteristics as transformations; that while "atomizing" people into qualities to concentrate on they take no consideration to lay out that the characteristics so atomized could quite develop by regular choice; that they overlook formative imperatives on development.

That they neglect to recognize attributes that are common because of causes other than normal determination; that they neglect to recognize current adaptiveness and a previous history of normal determination; that they create adaptationist speculations, neglect to test them appropriately and supplant one such speculation with another, disregarding other fascinating sorts of evolutionary and non-evolutionary clarification.

All things considered, as indicated by Gould and Lewontin, adaptationists tell simply theoretical, untestable "just so" stories and present them as scientific truth.

6.4. HUMAN BEHAVIORAL ECOLOGY

HBE is a relatively new and overtly Darwinian method of studying human behavioral variation. It addresses a lack of evolutionary principles in the social and behavioral sciences' examination of human behavior and social structure.

It has fostered the formation of new publications and academic specializations especially committed to preparing students to question the evolutionary origins of behavior since its start in the late 1970s. Its impact on academic life has been inconsistent. And it's most visible in fields such as anthropology and psychology, where human evolution is seen to be important for a comprehensive knowledge of human behavior, or in economics, where self-interest, methodological individualism, and maximizing are fundamental aspects in modeling behavior.

It has been met with resistance in areas like sociology and political science, where results on human evolution are either disregarded or seen as useless.

The second major modern method of using evolutionary theory in archaeology is HBE. Archaeologists working in this paradigm believe that humans (and other animals) have developed to seek success and make relatively rational judgments about things like optimizing the rate of food intake, defending resources, and so on, to the extent that they are related to Darwinian fitness.

They investigate the extent to which predictions based on these assumptions are supported by the archaeological record using methods developed in zoology and evolutionary ecology, which were heavily inspired by economic approaches.

Decision-making process is commonly thought to be based on the rational decisions among different models, in conjunction with individual learning and creativity, and in the context of limited and costly knowledge, as well as varied limits in feasible courses of action. HBE is similar to sociobiology, although that area focuses on reproductive success more directly through the use of kin selection and other models that explicitly evaluate Darwinian fitness, rather than evaluating nonreproductive behaviors assumed to be connected to health.

Natural selection has built creatures to respond to external situations in fitness-enhancing ways, according to the adaptationist approach in modern evolutionary biology. Using this as a starting point, behavioral ecologists develop and test formal models with particular optimization objectives, currencies, and restrictions, which they then use to examine the evolution and adaptive design of animal behavior in the ecological context.

Some anthropologists and cognate discipline researchers have used this methodology, together with theory and method from their home field, to investigate human behavior. HBE first appeared in the 1970s and expanded dramatically in the 1990s (Winterhalder and Smith, 2000). It is not as extreme a departure from traditional social science as it appears since it integrates information from the much older school of ecological anthropology and pays some attention to the roles of intentionality and cultural change.

Human behavioral ecologists must evaluate an extremely labile and causally complicated set of phenomena since they focus on behavior, particularly social behavior with a significant cultural component. They often try to explain such complicated patterns of behavioral variation as a phenotypic adaptation to changing social and ecological contexts.

The emphasis is on testing hypotheses about the relationship between environmental circumstances or payoffs and behavioral variance, rather than on the developmental or learning mechanisms that establish or sustain this match. The hypothesis that such phenotypic adaptation is directed by "choice principles" provides a link between it and genetic evolution.

These decision rules are thought to be panhuman cognitive adaptations that emerged through natural selection (or recurring cultural development) and govern behavioral decisions variation in ways that are responsive to environmental context.

Behavioral ecologists use relevant components of environmental variability to explain behavioral variability across time or within a population at a particular moment in time (broadly defined to include the social environment). Explaining changes in group size, sharing/exchange, land tenure/resource defense, prey choice/diet breadth, and settlement pattern are all traditional challenges.

Although these issue areas appear to be small at first look, they are clearly wide enough to allow academics in this discipline to investigate some of the most important changes in human history, such as the shift from foraging to agriculture and animal domestication. Furthermore, recent HBE models are progressively dealing with broader social and even symbolic issues.

The application of signaling theory, which proposes that self-interested individuals can be expected to engage in individually costly but frequently group-beneficial signaling behaviors, with such behaviors becoming more socially conspicuous in proportion to an individual's ability to produce them, is one such development.

According to Rebecca Bliege Bird and Eric Alden Smith, expensive (honest) signaling explains problems such as unconditional giving (for example, in feasting), the pursuit of dangerous large games for public consumption, conspicuous religious dedication, and creative excellence, massive building, and so forth.

6.4.1. Sociobiology as Human Behavioral Ecology

HBE, often known as human evolutionary ecology, is the modern evolutionary social science most closely connected to Wilson's sociobiological endeavor; it is the study that some philosophers of science still refer to as "sociobiology."

"Sociobiology" is most often used in this context to contrast with "evolutionary psychology," another current evolutionary social science project inspired by early sociobiologists and much developed from how early sociobiology was conceived: see the entry on evolutionary psychology for more information.

HBE is often known as "evolutionary anthropology" or "Darwinian ecological anthropology."(owing to the fact that the majority of its practitioners have a background in, or are located in, anthropology departments, and use anthropological research to test their assumptions about human behavior). The various labels are usually used when the researchers are stressing the linkages between their subject and anthropology or non-human behavioral ecology. HBE shares several similarities with the non-human behavioral ecology mentioned above. It is notable for its emphasis on behavior rather than the psychological mechanisms indicated in evolutionary psychology. In contrast to non-human behavioral ecology this has continued to use Grafen's (1984) phenotypic gambit, in that its practitioners seldom seek to uncover or characterized the genetic or other resources used in the creation of the behavior in question, nor do they generally attempt to identify the underlying psychology.

There are some changes in focus, however, in that the concerns regarding human behavior addressed by human behavioral ecologists (HBEs) differ from those addressed by non-human behavioral ecologists.

6.5. CENTRAL CRITICISMS OF SOCIOBIOLOGY AND BEHAVIORAL ECOLOGY

There are several philosophical objections to behavioral ecology and its methodologies; this section will examine some of these objections.

6.5.1. Adaptationism

Maybe the most serious criticism levied about "Pop Sociobiology" was that it was overly adaptationist; similar complaints are leveled at later successors of sociobiology, such as behavioral ecology.

Behavioral ecologists, including HBEs, are methodological adaptationists in the sense that they assume natural selection is optimizing the trait they are studying as a heuristic in order to establish other things about that trait; as a result, no consistent approach to the principle of natural choice is needed on their part (as discussed earlier).

Many behavioral ecologists saw Gould and Lewontin's (1979) objections as a cautionary tale rather than as legitimately intended at them—for example, Gould, and Lewontin's complaint that adaptationist ideas were untestable, which behavioral ecologists readily refuted. Parker and Maynard Smith (1990) reacted to Gould and Lewontin's criticism that adaptationists fail to account for phylogenetic and developmental restrictions (Figure 6.3).

Parker and Maynard Smith stated that they are considered—even when developmental limitations and tradeoffs are not explicitly represented in optimality models, they are frequently integrated into the strategy set or are somehow implicit in such models.

However, it is fair to say that Parker and Maynard Smith's study did not adequately address Gould and Lewontin's objection: Gould and Lewontin acknowledged that adaptationist models include constraints; their concern was that such constraints are used ad hoc to fill gaps in models and are not subjected to any external testing.

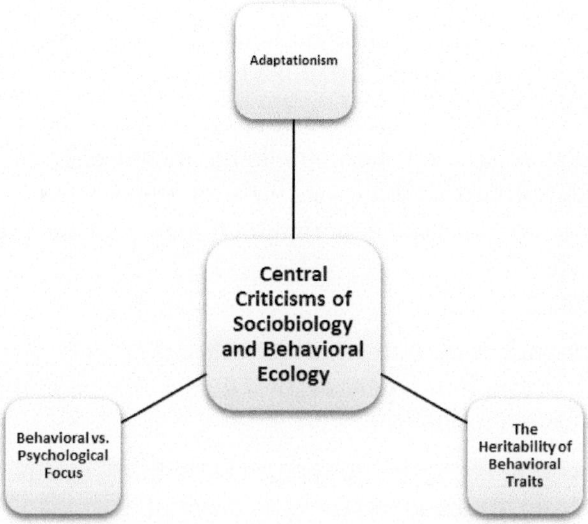

Figure 6.3. Central criticism of social–biology and behavioral ecology.

One more persistent criticism of Gould and Lewontin's adaptationist methods is that they do not allow scientists to consider non-adaptationist explanations for the traits to which they are applied, such as whether those traits were fixed by drift or are simply side effects of other traits (i.e., whether they are "spandrels").

In theory, adaptationist methods can help identify cases where a trait is a non-adaptation: drift, for example, can be identified by comparing the distribution of traits expected under selection and drift with actual population distributions (Sober, 2005); the question is whether the methods used in behavioral ecology are the most efficient at detecting these cases (Lewens, 2009).

Some philosophers argue that methodological adaptationism is more detrimental in practice than it looks in theory because true evolutionary social scientists are incapable of rejecting adaptationist hypotheses for a characteristic even when there is a compelling justification to do so.

6.5.2. The Heritability of Behavioral Traits

Whereas it appears that the allegation that perhaps the early "Pop" sociobiologists were genetic determinists was probably unjustified, there is a related objection that can be raised to both "Pop" sociobiology and later forms of human and non-human behavioral ecology, namely that any trait being an adaptation requires that the trait be heritable, and most human behaviors are not heritable.

To state a characteristic is heritable in the most basic sense is to suggest that if one parent possesses that trait, the trait is more likely to arise in the children (i.e., to say the trait is heritable simply indicates that the trait is inherited consistently).

The more stringent population-genetic definition of heritability states that a heritable trait T is one in which variation in T is predominantly accounted for by variation in genes rather than a change in the environment (see also the entry on heritability).

Natural selection is thought to require heritability in this sense because, for natural selection to spread a variant T in a population when T increases the average number of offspring of its possessors, those offspring must reliably also have T if T is to spread at the expense of the other variants. Unfortunately, many human behavioral qualities do not appear to be heritable in this way: they vary considerably between and even within cultures and settings, but human genetic variation is far too low to account for such changes.

HBEs, on the other hand, can get around this difficulty by pointing out that local patterns of overt behavior are manifestations of more general strategies; changes in the environment induce the manifestations of various parts of these complex behavioral dispositions that humans possess.

The hunting behaviors of the Ache people in the Paraguay rain forest and the Inuit in the Arctic, for example, are substantially different, and genetic variations between these groups are insufficient to explain them. Both, however, might just be local representations of a bigger and more universal foraging strategy (or strategies) that humans have, which could be heritable and an adjustment.

After all, a behavioral strategy is simply a complex behavioral disposition that involves responding to a set of environmental cues in a variety of specific ways, the majority of which would be achieved by having an appropriate psychological mechanism or mechanisms underlying the set of dispositions.

In such a situation, the heredity of a behavioral strategy is simply dependent on the existence of a psychological mechanism or mechanisms that consistently generate that strategy, and the heritability of that mechanism itself. Of course, whether such heritable processes exist for the behavioral patterns in which the HBEs are concerned is likely to be debatable.

6.5.3. Behavioral vs. Psychological Focus

The other issue with behavioral ecology and sociobiology is their emphasis on understanding the biological basis of behavior as such.

The behavioral focus made sense to those behavioral ecologists influenced by Lorenz and Tinbergen's ethology, but once cognitive psychologists became interested in applying evolutionary thinking to their work, the effects of the cognitive revolution in psychology began to have an effect on some of the "sociobiological" thinkers, and sociobiology was transformed into evolutionary psychology.

As a result of the cognitive revolution in psychology, many psychologists started to see truly psychological states or mechanisms as the primary object of explanation in psychology rather than behavioral tendencies (as in Skinnerian Behaviorism).

Since the middle of the twentieth century, cognitive science has primarily defined psychological states and mechanisms in computational and materialist terms, that is, that the mind is the brain, and that brain states and mechanisms are fundamentally computational states and computers, respectively (though almost certainly not serial computers).

Psychological descriptions, then, are merely computational or information-processing descriptions of these brain systems; usually, they refer to representations, decision rules, and algorithms that brain systems process. This approach adjustment constituted a fundamental shift in methodology and thought. However, it appears to have made evolutionary cognitive psychologists wary of seeing behavioral dispositions as legitimate goals of evolutionary explanation, just as they were wary of considering them as proper targets of psychological explanation.

This issue does not appear to have arisen for human behavioral ecologists; indeed, not only is behavior treated as the sole proper target of evolutionary explanation but psychology is explicitly ignored in favor of behavior for methodological reasons—the focus on behavior is part of the "phenotypic gambit" described above, which simplifies evolutionary modeling and the process of deriving explanations from it in HBE.

6.6. WHY ECOLOGY NEEDS BEHAVIORAL ECOLOGY

Many ecological results are the result of behavioral processes; hence behavior is crucial for ecology. Behavior determines how creatures interact with their surroundings. The predator-prey interaction is produced by behavior: a predator catches, discovers, and consumes its prey, while the prey avoids the predator, at least frequently enough to remain.

To understand this relationship, the ecologist collects data on predators and prey in order to identify the function that connects the two numbers, known as the functional response. The quantities of predators and prey are determined by behavior, namely how predators identify their victims and how prey hide or flee. Mutualistic interactions are produced by behavior.

An ant colony dwells in a hollow domatium in a tree limb, consuming the honeydew of scale insects that feed on the tree and guarding it against herbivores. The expansion of the ant colony impacts its behavior, particularly the behavior involved in monitoring the tree and fighting herbivores. Scale density, which restricts tree development, influences the efficiency of the colony's defensive activity, which encourages tree growth.

The behavior of ants is the connection that creates functional relationships between the tree, ant, and scale populations. In general, conduct creates ecological interactions, or, to put it another way, ecological relationships manifest themselves via behavior.

The manner in which behavioral ecology might contribute to ecology, particularly community ecology, revolve around the context-dependent themes of variation, change, dynamics, and history. Community ecology has struggled with the fact that sweeping generalizations are constantly weakened by context dependency throughout the last ten years.

"Context dependency has hampered the creation of a universal framework for anticipating the kind and amount of (risk effects," writes Heithus et al. (2009). I put quotation marks around "risk effects" because, while this assertion is about risk effects, the same may be stated about the difficulty of generalizing about practically every ecological consequence.

As Lawton (2000) points out, ecological results for a specific species in a specific location seldom provide general principles. Because environmental change causes ecological and evolutionary transformations, "the local rules of interaction within a community must likewise change, typically gradually but occasionally abruptly." As a result, hard-won local insights from community ecology research are basically fleeting." Ecologists have progressed more in their knowledge of the significance of behavior to ecology than behavioral ecologists have in their grasp of the importance of ecology to evolutionary problems regarding behavior. Ecological journals are more likely to accept manuscripts on behavior than behavioral ecology journals are to accept articles on ecology.

Ecology has devoted special issues to behavioral subjects such as trait-mediated interactions; Oecologia lists "behavioral ecology" as a category in a list that encompasses everything from physiological to global ecology. Behavioral ecology publications, on the other hand, seldom publish work that demonstrates how environmental conditions impact the implications of behavior for reproductive success.

It is time to admit that behavioral ecology and ecology are not mutually exclusive and to merge behavioral ecology and ecology. Behavioral ecology and ecology share projects. Community ecology can give behavioral ecology the tools it needs to raise genuine questions about natural selection's present operation.

The question in evolutionary ecology has shifted from "Why does characteristic x contribute to reproductive success?" to "What are the conditions in which trait x contributes to reproductive success?" This ecological viewpoint must be applied to the study of behavioral evolution. Community ecologists have realized that human activity has an impact on ecological results.

For example, behavior influences the effect of history on community assembly, indirect effects in predator-prey interactions, and population reactions to human disturbance. Quite broadly, conduct is frequently the cause of context dependency; various behavioral responses in different contexts result in diverse ecological results.

As the field of community ecology expands to encompass behavior, behavioral ecologists can begin to use ecological viewpoints when posing evolutionary issues.

Behavioral biology and sociobiology are terms that assign a way to deal with the investigation of behavior in which request ordinarily centers around how behavior has developed by regular determination, particularly how behavior contrasts among species have emerged corresponding to their common habitats. As sociobiology, which is restricted to a social way of behaving, is basically included inside the term conduct environment, the custom has come to incline toward utilization of the more comprehensive term behavior ecology, The predominant integrative subject of behavioral ecology, to be specific the job of regular determination, is tragically not unequivocal in both of these terms. In any case, both include the nearby reconciliation of thoughts from environment, evolution, and ethology.

6.7. CONCLUSION

Behavior of organism's links physiological and molecular genetic systems with the external environment. The behavioral study of behavior is at the heart of how evolution operates in nature. Behavioral ecologists attempt to explain the functional evolution of behaviors in all types of organisms, not just animals. Most modern projects studying the evolution of behavior seem to have developed theoretical and methodological features of their own. These projects are sometimes referred to as the "evolutionary social sciences."

The section will briefly describe the relatively uncontroversial non-human animal behavioral ecology that Wilson described as "socio-biology." Behavior is what animals do and why they do it. There are two broad categories of behavior—learned and instinctive. Instinctive behavior is a pattern passed down from one generation to the next. Learned behavior is the result of experience accumulated and assimilated throughout a lifetime.

REFERENCES

1. Davies, N., Krebs, J., & West, S., (2012). *An Introduction to Behavioral Ecology* (4th edn.). [e-Book] A John Wiley & Sons, Ltd., Publication. Available at: http://www.reihanedu.com/wp-content/uploads/2020/11/NB-Davies_-J-R-Krebs_CompressPdf.pdf (accessed on 27 June 2022).
2. Gordon, D., (2011). The fusion of behavioral ecology and ecology. *Behavioral Ecology, 22*(2), 225–230. [online] Available at: https://academic.oup.com/beheco/article/22/2/225/206866#2173338 (accessed on 27 June 2022).
3. Plato.stanford.edu. (2013). *Sociobiology (Stanford Encyclopedia of Philosophy)*. [online] Available at: https://plato.stanford.edu/entries/sociobiology/#SocBehEcohttps://www.oxfordbibliographies.com/view/document/obo-9780199941728/obo-9780199941728-0015.xml (accessed on 27 June 2022).
4. Royle, N., (2013). An introduction to behavioral ecology. *Animal Behavior, 85*(3), 686, 687. [online] Available at: https://www.sciencedirect.com/science/article/abs/pii/S0003347213000122?via%3Dihub (accessed on 27 June 2022).
5. Smith, E., (2003). *Human Cooperation Perspectives from Behavioral Ecology*. [e-Book] https://faculty.washington.edu. Available at: https://faculty.washington.edu/easmith/dahlem-ch.pdf (accessed on 27 June 2022).

CHAPTER 7

Biological Species As a Form of Existence, the Higher Form

CONTENTS

This chapter begins with elaboration on the problem encountered while defining the self-producing unit in sexual reproduction. Simultaneously, deme or the subdivision of population that contains individuals closely related to each other is also highlighted. Multiple theories regarding sexual reproduction and genes are discussed in a challenging way that dismisses their credibility while lauding some.

Additionally, mechanisms of gene pertaining to its essential process-replication are also stated. Behavior of the gene is discussed in a way that equates it with the macroscopic individual in terms of being a selfish gene or an altruistic gene. The philosophical dimensions of existence of self-reproducing entities and their significance at large in the grand scheme of universe is touched upon.

Finally acknowledging the immense source of information organisms carry within themselves in the form of genes, the chapter calls for a deeper understanding of the finesse individuals possess and need to work upon through the aid of existing knowledge.

7.1. INTRODUCTION

The Universe is made up of distinct things such as elementary particles, atoms, molecules, planets, stars, and galaxies. That is, there are just a few matter configurations that are relatively stable and long-lasting, with the intermediate ones being volatile. The Universe is organized. It signifies that the world's components, as well as the Universe itself, are resistant to disorder.

They are a long way from reaching thermodynamic equilibrium. They do exist. Existence is a form of resistance to chaos. Many people believe that thermodynamic equilibrium signifies death for biological entities, however this is true for any system. It may seem strange now, but Darwinian natural selection was originally thought to establish an equal continuum of living organisms by accumulating microscopic heritable modifications.

This prediction was never fulfilled. The absence of this continuum is still used by creationists to argue against biological evolution. The biological world, on the other hand, adheres to the same global principle: individuals, populations, and species are distinct stable entities, with intermediate configurations being dynamic.

Biological evolution cannot and does not maintain everything that develops at random. The Universe's existence is dependent on the mutual affinity of its parts, their capacity to interact with one another, therefore rejecting the universal need for evenness. However, this is merely one side of the coin. The contact should keep the dissipation to a minimum.

Any life necessitates a balance of two opposing forces–dispersing and compressing, repulsion, and attraction. Too much contact causes collapse and disappearance in singularity. There are two types of existence: animate and inanimate. Any existence does not last forever. The second rule of thermodynamics states that everything in the Universe will eventually dissipate or collapse.

What is the force of dissipation? In general, it refers to energy. Why doesn't the Universe just fade away? The Universe's components interact with one another, slowing dissipation. The four fundamental physical interactions (weak and strong nuclear interactions, electromagnetic interaction, and gravity) prevent the inanimate things from dissipating immediately.

These interactions generate a plethora of new forms of existence, new creatures. The entire Universe is evolving. The major principle of prebiotic evolution is the same as that of biological evolution. Various things constantly originating in the Universe have varying lifetimes ranging from fractions of a second to billions of years. Ephemeral forms are replaced by more permanent ones during evolution.

This idea–"survival of the fittest"–appears to be a tautology, but it is the supreme tautology: Everything really novel arises from this premise. "Persistence is the only relevant and objective definition of adaptation." Longevity is a numerical measure of life.

As a result, the Universe's evolution entails the development of resistance to dissipation, disorder, and entropy. Longevity, on the other hand, indicates a restriction of the force of interaction, resistance to collapse, and annihilation in singularity. To survive, living things must be able to maintain a balance between dissipation and attraction.

7.2. BIOLOGY IS SPECIAL

It is becoming clear that life is not a physical process. It is more than simply advanced physics and chemistry. The higher kind of existence is an entirely distinct form of being. The swallow builds its nest in my shed outside Moscow, then flies to South Africa and returns year after year. Nobody has attempted to quantify the likelihood of such an occurrence arising from physical causation and stochasticity. It is clear that swallow knows the way there and back. I also know how to go from my house to work and back. And this information is simply a small percentage of the overall knowledge that each creature possesses. This entire knowledge is the ability to recreate itself. Please excuse those who can't bear anthropomorphisms.

Humans are biological creatures with numerous distinguishing qualities in common with other living entities (and vice versa). One of them is knowledge. Knowledge is defined as the capacity to do a low-probability action consistently. The knowledge may be expressed quantitatively as the inverse of the action's probability (Figure 7.1).

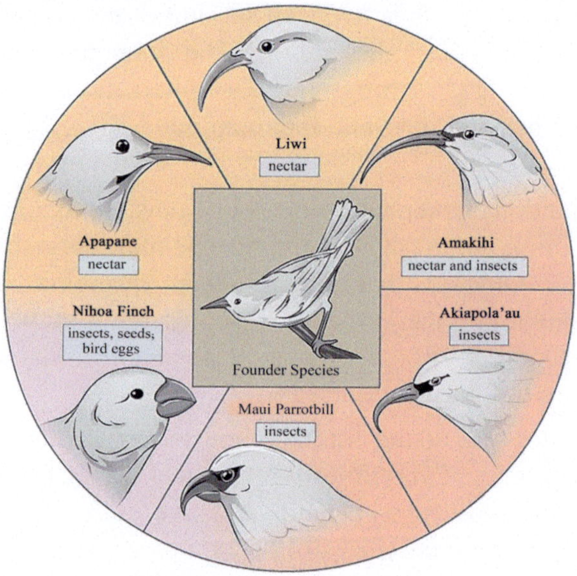

Figure 7.1. Image showing founder species.

Source: Image by Wikimedia commons.

This skill develops throughout species evolution and organism ontogenesis (not only in a high school). Knowledge is connected to conduct when defined in this way. A knowledge-based conduct is an expedient behavior. Using the term "behavior" broadly, we may refer to the behavior of molecules, organelles, cells, organisms, populations, and ecosystems.

Physics' dictatorship stops at the boundary between the inanimate world and the biosphere, when the realm of sense and knowledge starts and matter's behavior becomes expedient. Knowledge, memory, coding, transcription, translation, function, signaling, recognition, decision-making, regulating, and creation become not just acceptable but unavoidable when defining living systems.

7.3. ORGANIZATION AS A FORM OF EXISTENCE

An organization is a complicated system that can execute certain duties due to the unique arrangement of its pieces. Systems that are organized must be separated from those that are ordered. Both systems are random, but the ordered systems are formed using a simple method and so lack complexity, whereas the structured systems must be constructed element by element using an external programmed or design.

Organization is complexity with a purpose. It is not random because of design or selection, but rather because of the requirement of crystallographic order. Only living things (cells, organisms, populations, and species) are considered. Machines are also organizations, but they are human-made animals with the function of survival. Parts of the biological creature may only perform specific duties.

"Life is built on semiosis, or signals and codes," and physics and chemistry cannot properly describe it. Everything important in biology is decided by semantic norms and goal-directed programming rather than physical causes. This idea is applicable at all levels of biological structure. The coding of polypeptide sequences using nucleotide sequences is not restricted.

The complete life cycle is carried out by the progressive generation of organic codes and interpretation rules for the entity's gradual self-manufacturing. In contrast to the inanimate world's objects, which are created by random interactions, living organisms and their components are created through ontogenetic purpose. "All biological objects are artifacts, and … life is artifact-making."

Symbols are used in coding and instructions, but they are tied to the represented topic conceptually rather than physically. The DNA sequence cannot be derived from the nucleotide physic-chemistry, just as the text cannot be determined from the alphabet. Furthermore, non-deductibility is a need for DNA's capacity to code genetic information.

The objective and promise of molecular biology were to explain biological phenomena using highly sophisticated physics, namely non-equilibrium thermodynamics. The inadequacy of this method to explaining life becomes clearer by the day. Living things live not because of any unique physics or chemistry, but because of their rational conduct. The Universe's full content is represented by two interconvertible essences: energy and substance or matter. Recently, a group of daring individuals began discussing the third essential element–biological information, which is neither energy nor matter. It is a characteristic of life and only life. The term "information" may not be the best choice.

It is similar to the homonym used in information theory. Shannon's information is meaningless, yet meaning is precisely what we are interested in the context of biology. Most crucially, "the third essential essence" is behavior, which involves internal and external communication, interpretation, and implementation in the form of structure with a survival purpose.

"Mind" is an appropriate word for this third essential essence. Four fundamental physical interactions keep the inanimate universe from dissipating immediately, but they do not explain the presence of living organisms. The living world uses behavior to combat chaos. Let us define reasonable conduct as behavior that is targeted at ensuring the survival of the acting organism.

Inanimate evolution might be described as Nature's self-construction: matter from energy; biological evolution could be described as nature's self-knowledge: mind from matter. The mind is more than a byproduct of existence. It's the fifth essential interaction that keeps the living world from disintegrating and collapsing. The mind is an important life-specific survival tool. The terms "thought," "knowledge," and "sense" should not be interpreted metaphorically.

These are excellent terms. My body's cells understand how to copy DNA. This information is theoretically similar to my ability to read and write, but it is obviously far more crucial for my existence than literacy. The whole knowledge possessed by a living thing is the ability to replicate itself. Neither more nor less. Life is endless self-replication.

All activities in inanimate nature are oriented from a less likely condition to a more probable state: movement to equilibrium. Life is a journey toward an improbable condition. It is a river that is rising. A physical process is a stone falling from a mountain; a biological process is an alpinist going up a mountain. It requires not just energy but also intention (will), knowledge, and consciousness (Figure 7.2).

During this "climbing up," no physical rules are broken. Instead, life's goal-directed algorithms harness them in such a manner that low-probability, nearly impossible events become the most likely. As a result, life is a distinct form of existence from all other forms of inanimate existence. A live being might vanish for one of two reasons: death or transformation.

Figure 7.2. Image showing a biological species–honeybee.

Source: Image by Wikimedia commons.

In both circumstances, the prior creature is no longer alive. Existing means staying alive for a long time. Biological creatures know how to survive for a long time. Because living things are incredibly complicated, this understanding is miraculous. The fact that biological complexity is well ordered is remarkable, not the complexity itself.

The term "organized" suggests that only a few numbers of states are compatible with vitality, but "complexity" means that the system has a huge array of states to choose from. "For the route to life is narrow and the entrance is difficult." Living systems, in other words, are non-equilibrium, low-entropy, and low-probable entities. The second law of thermodynamics is not broken by creatures.

Living organisms are open systems that exchange matter and energy with their surroundings on a regular basis. Organisms release entropy into the environment, keeping their internal states out of balance. This is an accurate description of the situation, yet it falls short of explaining biological organization. For the execution of life, energy is necessary yet insufficient. It necessitates knowledge.

Energy is a chaotic force in and of itself. Energy must be wisely harnessed to support organization. Living things live because of their rational behavior. They are aware of the path that leads to life. By the way, the importance of knowledge for the application of life was already apparent to the eighteenth-century nature philosophers.

7.4. LIVING ENTITIES

7.4.1. Prokaryotic Cell

A cell is the most basic form of biological existence. A prokaryotic cell is the most basic biological unit. Though any independent thing in English can be called an "entity," it is heuristically necessary to name only the living or biological entity that has the function of survival. Proteins, nucleic acids, membranes, ribosomes, viruses, and other cell components have specific roles and are hence not biological beings (Figure 7.3).

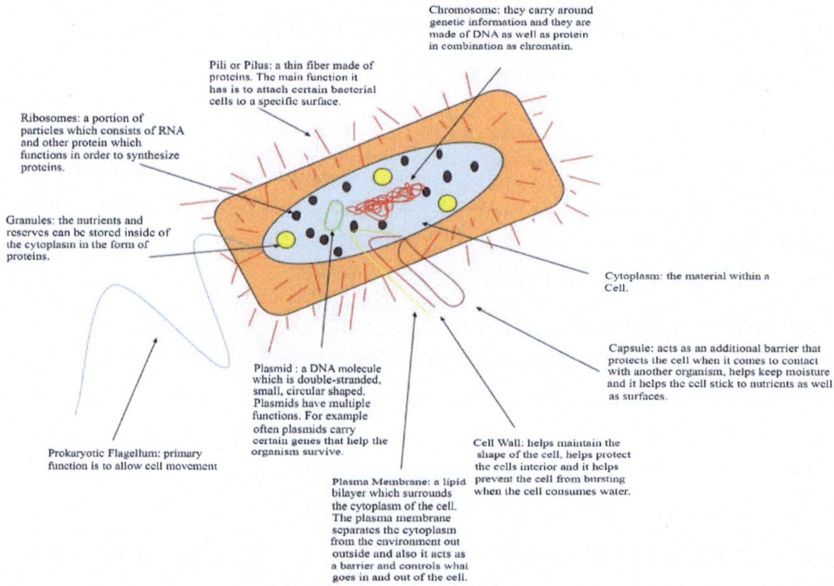

Figure 7.3. All organelles and components of cells related to genetic material (DNA/RNA).

Source: Image by Wikimedia commons.

A monad is a living creature. It is isolated from its surroundings and forms a cohesive whole; it exists in time through self-reproduction. Possible problems with such a definition of a living being have been noted; however, they are not only useful but also inevitable in the context of this study.

A cell is the smallest unit of Life. A live being cannot be smaller than a single cell. Cellular activity produces the components indicated above. Prokaryotic cells are also the oldest living things on the planet. They first appeared three billion years ago on Earth.

7.4.2. Eukaryotic Cell

Eukaryotic cells are far more complicated. It first appeared on Earth approximately one billion years ago as a result of cooperation between many prokaryotic cells. As a result, the vast and diverse world of unicellular eukaryotes was born (Figure 7.4).

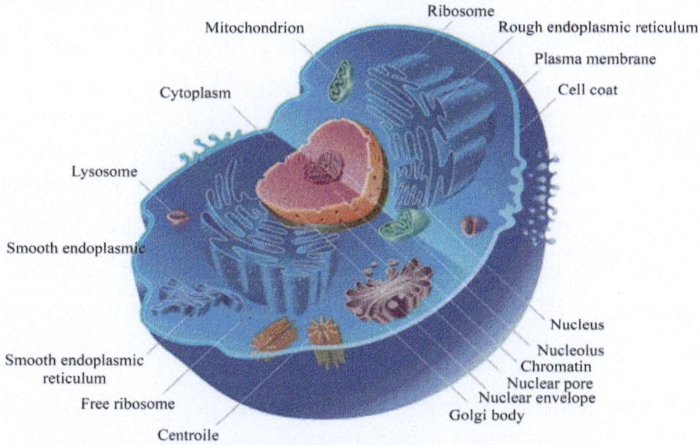

Figure 7.4. A typical animal cell with labeled organelles.

Source: Image by Wikimedia commons.

7.4.3. Multicellular Organisms

Eukaryotic cells' greater complexity allowed for more cooperation and the emergence of multicellular creatures. A monad is a multicellular organism. It is a substantive entity (see below) that can replicate itself acting alone if it is asexual.

7.4.4. Biological Species

Higher multicellular creatures with huge genomes and intricate development have been found to be unable to reproduce themselves consistently over generations. This stumbling block prompted even further collaboration: the construction of a multi-organismic entity that could reproduce itself–a biological species. Individual organisms within the species do not have substantial monad status.

The approach presented above is simplified. On the one hand, a generation of interbreeding population–deme–is the true self-reproducing entity. On the other hand, there are many different kinds of sexuality, not just the necessary kind. The compulsory sexual forms communicate the existential repercussions of sexuality the most plainly.

7.4.5. Substantive and Attributive Existence

In accordance with the above definition of a living or biological being, the terms "substantive" and "attributive" existence will be used here. In terms of reproduction and evolution, substantive existence means the entity's autonomy and self-sufficiency. On the theater of life, a substantive entity is a sovereign player.

Attributive existence denotes the presence of a higher-ranking entity (host). Its survival and evolution are causally connected to the host's survival and evolution. An organism is a substantive entity in the case of asexuality, whereas a sexual individual is an attributive entity that is a part of a higher rank entity–deme.

Attributive existence is the most common type of biological existence. Hepatocytes, for example, are entities that exist as a feature of mammalian species. They belong to a group of polyphyletic organisms. Hepatocytes from various species are far more similar to one another than, say, neurons or other cells from the same mammal.

They multiply and evolve as an attribute of the host and for the benefit of the host while displaying trans-specific "epigenetic consanguinity." Organelles are another wonderful example of attributive existence (mitochondria, plastids).

The concept of substantive and attributive existence has been shown to be useful in the description and comprehension of biological organization and evolution. It is critical to remember that the existence of a sexual individual organism is attributive in the context of this book.

7.5. EXTANT AND LASTING FORMS OF BIOLOGICAL EXISTENCE

As previously stated, as the Universe evolves, transitory forms are being replaced by more permanent forms. In our ever-changing environment, this outcome is unavoidable. Inanimate entities are the most likely configurations of matter in a given scenario; their physical durability, which is provided by the balance of dissipation and attractive power, ensures their longevity. Organisms are physically frail, immensely complicated, low-entropy systems with low probability combinations of matter. They can't keep up with entropy growth indefinitely. The homeostatic systems will never be perfect. They make mistakes and lose their toughness. Perfect equilibrium would necessitate endless energy expenditures.

Even in the most advantageous environment, in the absence of any competition, and with an excess of energy and substance, organisms die. Entropy causes them to perish. It appears that organisms have reached their thermodynamic limit and are unable to develop their homeostatic capability any more. The accuracy of the cell processes has been fine-tuned to perfection.

Both too little and too much precision is detrimental to organismal viability. The focus of energy expenditures is on DNA replication fidelity. And it's quite high: one wrong nucleotide is only integrated once every 108–1010 occurrences. With misincorporation rates of 1 in 104 and 1 in 103–104, respectively, transcription, and translation have a substantially lower fidelity.

A considerable fraction of newly manufactured polypeptides has amino acid alterations due to this mistake rate. And it isn't the only issue. Every biopolymer and supramolecular structure are constantly degraded, and the faults accumulate over time. Error accumulation must have self-accelerating dynamics that inevitably lead to disaster. By reproducing, living systems can avoid the catastrophe.

To prevent mortality, they reproduce. Single-cell organisms are sometimes referred to as immortal. It's a miscommunication. They also reproduce to save themselves. Even with Escherichia coli cells that appear to divide symmetrically, the two supposedly identical cells formed during division are functionally asymmetric.

The old pole cell should be viewed as an aged parent who produces revitalized offspring on a regular basis. Life's dynamic stability is one of its distinguishing features. Throughout their lives, living things change. This transformation is essentially self-regeneration, self-manufacturing, and self-renewal. This is what life is about.

Throughout time, organisms, and generations of species continue to reproduce. They are temporary, renewable forms of life. Lineages and species are the most durable forms of biological existence. A transient link in the existence of a lineage or species is an individual organism or a generation of species. Multiplication and reproduction may be combined to increase the number of organisms. This expansion is critical to the species' existence, but it is not guaranteed. The replacement of an old, worn-out body with a new one is the core of reproduction. It may not, and in most cases should not, lead to an increase in the number of organisms in order to avoid resource exhaustion.

Stable reproduction is the true measure of evolutionary success. Why does reproduction, a more complex phenomena than the individual's simple life, prevent entropy from increasing? The idea is that selection is always accompanied with reproduction. Natural selection is a form of reproduction quality control. Imperfect copies are discarded, whereas novelty may be retained.

This is the process of memorizing new information. Only if they improve, or at the very least do not significantly deteriorate, homeostasis. As a result, natural selection has mostly produced conservative results. Without conservatism, life would be impossible. The majority of evolutionary scientists focused on the generative, innovative aspect of evolution.

However, evolution is a two-fold phenomenon: it both creates and stabilizes innovations. Furthermore, evolution is merely a result of faulty reproduction. The changes are entropy-driven and do not appear to be mysterious. They are, rather, unavoidable. Stability, perseverance, and resistance to not only destruction but even subsequent change are truly marvels.

Only conservative mechanisms are linked to rational behavior (by definition). Reproduction along with selection is the only method capable of ensuring a living entity's long-term existence and potential immortality. Natural selection, it has been discovered, is not a particular goal-directed mechanism developed for lineage survival.

It works automatically by keeping those who persevere and releasing those who give up. Special mechanisms that improve the efficacy of purifying selection are recognized, however. In an ideal world, reproduction would be exact; otherwise, the aim of immortality would not be achieved. Template-directed synthesis was nature's first and most important invention, from which life arose.

It is obvious that DNA replication must be precise enough that the majority of progeny acquire undamaged genetic information. The true fidelity of DNA replication is astounding. For unicellular creatures, the achieved fidelity of DNA replication is sufficient for lineage immortality.

The number of mutations per genome every generation in multicellular organisms with huge genomes and complex growth is unacceptably high, up to three orders of magnitude higher than in yeast, for example. In other words, the genomes of higher creatures are not accurately replicated. In humans, for example, there are 60 or even more mutations per zygote.

Because of mutational overload, or entropy, higher species must rapidly degenerate at such a rate of mutation. They persist, however. Their existence necessitates an explanation. Why hasn't replication fidelity progressed to a greater level? The problem is that accurate reproduction is an expensive procedure. High precision necessitates an excessive amount of energy.

It appears that there was no way to improve genome replication fidelity any further. As a result, higher creatures must be able to effectively complete ontogenesis, and species must be able to endure throughout time despite constant mutational disturbances. In the context of asexual ("homeo-genomic") lineages, this dilemma has no solution.

During the irresponsible evolution, they would swiftly degenerate and become extinct or blurred. Various than high fidelity genome replication and purifying selection, evolution devised other strategies to avert or avoid the deadly outcome of the mutational flood, as previously discussed. The problem is intimately related to the occurrence of canalization or robustness.

Robustness is a feature that permits a system to sustain its functions in the face of external and internal disturbances. It is the ability to undertake successful ontogenesis despite environmental and mutational disturbances in biological systems. Robustness is defined as the ability to maintain function (meaning) despite structural and environmental changes.

All layers of biological organization contribute to robustness resources. One of the criteria of canalization is the global degeneration of the relationship between structure and function: there are more genotypes than phenotypes. During evolution, function is chosen over structure. It's possible that different genotypes correspond to the same phenotype.

This notion applies to several levels of biological organization, including the activation of multiple paths that lead to the same end outcome. This is conceivable because biological processes are governed by semantic rules and goal-directed algorithms rather than physical causality.

Simple creatures have simple semantics and hermeneutics because they reproduce their genomes with high fidelity. Higher organisms' languages are far more complicated, with extensive synonymy and a complex, context-dependent hermeneutics. This aids in development resilience to mutations and environmental disturbances.

Nonetheless, we must confess that all of these beneficial attempts have proven insufficient for higher species with big genomes and intricate development: they fail to reproduce themselves reliably through generations. Another instrument was required for their longevity. This instrument was sexual reproduction, or the construction of a self-reproducing multi-organismic entity.

For a long time, evolutionists paid little attention to the issue of sexual reproduction's emergence and persistence. The situation was perhaps too evident. No one questioned Weismann's theory that sexual reproduction promotes genetic variety, which provides material for natural selection and increases the species' evolutionary potential.

Fisher and H. J. Muller quantitatively established the likelihood of amphimixis accelerating evolution. Evolvability is still a prominent notion among population geneticists. It is commonly considered that the ability to evolve quickly and diversely is a good quality supported by natural selection, and that a lack of evolutionary potential is fatal.

The concept of evolvability as a selected attribute runs counter to established evolutionary efforts to maintain genetic stability in organisms and lineages. It is clear that evolvability cannot be considered a species homeostatic mechanism. Because direct selection for evolvability is logically impossible, the move to sexuality requires an explanation apart from evolvability.

Though sexual reproduction and genetic recombination are sources of combinative variation in populations, they simply produce new combinations of existing alleles, which are condemned to be deleted in the next generation. If one believes that sexual reproduction was established to speed up evolution (a Lamarckian viewpoint), then the continuous swapping of genomes (regardless of their quality) appears weird.

We should not presume that evolution has generated particular methods to accelerate evolution. Suicidal mechanisms, in this case. A species that evolved quickly would not last long. All of the organisms that live on our planet now are members of species that are resistant to further evolution.

Evolution is unavoidable because the systems developed by evolution to protect against evolution, species homeostasis, are not perfect and are occasionally overcome by entropy pressure. All species are capable of evolution simply because they descended from ancestors who were able to evolve and acquired their flaws, or "original sins." Evolution is difficult to avoid.

7.6. BIOLOGICAL ENTITIES AS SELF-REPRODUCING UNITS

A certain problem arises in delineating the self-reproducing unit in the case of sexual reproduction. Two individuals belonging to different sexes are enough to produce progeny. However, it is known that the stable reproduction requires a rather large interbreeding population–deme. Low robustness is found in small populations because of inbreeding that results in considerable homozygotization.

Generally, individuals that are homozygous have a drastically reduced vitality, and the populations formed by them also have low robustness due to the of lack of polymorphism and weak genotypic plasticity. A species, on the other hand, may consist of many demes with rare inter-deme genetic exchanges due to geographical constraints or habitual preferences.

Intraspecies organization and sexual relationships differ between species. As a result, the boundaries of a self-replicating unit must necessarily be ambiguous. This complication does not appear to be significant in the context of this paper. At any form of sexual relationship, all of the existential benefits of sexual reproduction are fully realized on the level of deme.

Rare inter-deme genetic exchanges may complicate the picture for evolutionary theorists, but they do not fundamentally alter it. Though demes are not completely closed entities, they are close enough to rely on their own merits for survival. The early group selection models were flawed since they assumed that genes acted independently, whereas currently it is apparent that gene interaction, and more importantly, interactions that are genetically based among individuals, were an important source of the response to group selection. As a consequence, many people are starting to recognize group selection as a potentially important force in evolution.

So, try to avoid the painful discussion about group selection here. The statement that a unit of selection and a unit of substantive reproduction are purely the same units, the same monads, has been observed. The notorious replicator/interactor discrimination was invented to save the erroneous selfish gene theory.

The evolutionary theory contains two major delusions: the gene as an ideal replicator and the individual organism as the quintessential unit of selection. Gene is not a physical entity. It has been observed that it is not an entity on any level because it is not an organization. It even lacks the distinguishing characteristics of an autonomous thing (Figure 7.5).

Figure 7.5. A pair of bettas spawning under a bubble nest.

Source: Image by Wikimedia commons.

It is a product of cell activity, even if the very important product. Functionally, It is a piece of text that acquires meaning only in the context of the entire organism. It is becoming clear that the concept of a selfish gene is not grounded in reality. The link "one gene–one trait–one selection vector" is not observed: one gene can affect numerous traits, and most traits are influenced by multiple genes.

A sexual individual, on the other hand, is not a self-replicating entity and thus cannot be selected as such. Organisms are distinct, one-of-a-kind components of species produced as individual units during species reproduction. Through inclusive fitness, the selective meaning of an individual organism is appreciated in the context of a population.

Deme generation is a small entity that reliably reproduces itself with high fidelity (according to the Hardy-Weinberg law). Ontologically, deme, and species are comparable entities. They only differ in size and degree of cohesiveness. As a result, we can say that a species' generation is a self-reproducing unit. In most cases, "deme" and "species" are interchangeable.

Thus, a substantive sexual entity is a deme or species generation. The term "generation" may require certain clarification. Deme (and species) is a lasting form of existence; generation is a transient (extant) form of a species existence. Now and here only generation will be dealt with.

As per observations it has been observed that, it is heuristically useful to keep in mind that actually living spatiotemporally restricted entity is a generation, not a lineage, which is in fact a historical phenomenon. A lineage could even be thought of as existing only in our minds. However, there must be something more substantial, something more existential in the final existence than our human perception of historical reality "The defining characteristic of a living organism is that it is the transient material support of an organization." This definition is fully applicable to a species generation as well. The material support is transient. But what is permanent?

According to Merser, organization is lasting. I.e., mind is lasting. The knowledge how to reproduce itself is lasting. Exactly and only this knowledge is transmitted from generation to generation. And of course, this knowledge is not a DNA sequence, singularly. Only entire substantive entity is a carrier of this entire knowledge, not only genes and brains.

It has been observed that, the "material support" and the mind are not separate things like the hardware and software of a computer (organism is not a turing machine). They formulate the aspects of the same whole. A real extant population may include organisms of varying ages, ranging from newborn to mature to elderly (overlapping generation), or it may be more or less synchronous.

A generation's lifespan is equivalent to the average lifespan of the organisms that comprise the generation. In the case of asexual reproduction, the existing entity is an individual organism, and in the case of sexual reproduction, it is a generation of deme.

When we say a species lives a million years, we may mean that an entity very similar to the extant entity lived a million years ago and is directly connected to the extant entity via a series of reproductions.

7.7. MEANING OF SEXUAL REPRODUCTION

Sexual organisms are subsets of a higher-ranking entity–biological species. The transition to sexuality, like all major evolutionary transitions, is a cooperative process. Individual organisms have lost their ability to reproduce and evolve autonomously. They exist and evolve in biological species. Their existence is attributive.

The transition to sexuality is ascension to a new and a higher quality. Sexual population is a self-reproducing, coherent system. Species reproduction is not the same as speciation. Species reproduction does not result in the formation of another (daughter) species. Reproduction is how species survive. The reproduction must be accurate.

In the case of a stable species, reproduction is extremely precise. Because of a species' genotypic plasticity and the organisms' phenotypic plasticity, one generation may differ slightly from another in response to environmental variations. These modifications are reversible manifestations of species rigidity. They are not changes in evolution.

Speciation, by the same token, is not species reproduction, nor is it the continuation of the given species. Speciation is a macroevolutionary event that is distinct from reproduction. Essentially reproduction is a conservative process. While reproduction is the renewal of the same, speciation on the contrary, is the creation of the new.

A species, like a sexual organism, is unique because it appears only once in the history of the biosphere and then disappears. As a result, it is incorrect to regard a species as a segment of a species level lineage, as De Queiroz suggests. There are no "species level lineages" in sexuality, just as there are no "organism level lineages."

A lineage is a chain of generations. It represents species ontogenesis in its entirety. Speciation is not a predetermined stage in the evolution of a species. To put it another way, the species is never "interested" in speciation. For the extant species, begetting a "daughter" species means begetting a competitor. It should be noted that allopatric is a requirement for the survival of a newly born species.

Reproduction of asexual and sexual organisms differs in many respects. Asexual organism (typically it is a cell) is a self-reproductive unit by itself. Its ontogenesis is commonly straightforward, moving from the young cell to the mature cell. The reproduction is flawless. It is accurate and reliable. Prokaryotic lineages appear to have existed for billions of years.

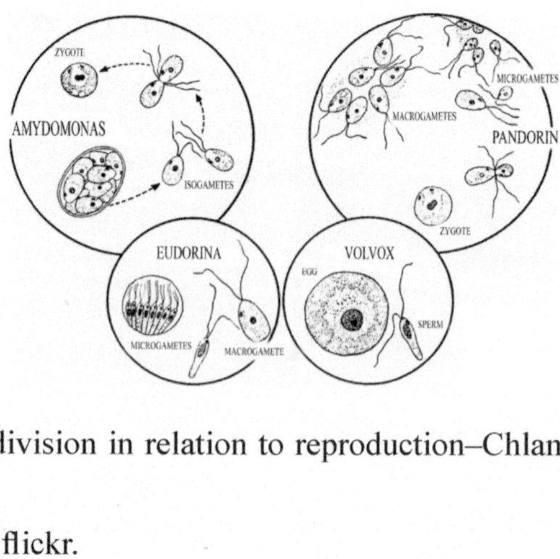

Figure 7.6. Cell division in relation to reproduction–Chlamydomonas Pandorina.

Source: Image by flickr.

The common (i.e., reductionistic) explanation attributes this incredible longevity to the high fidelity of DNA replication. A prokaryotic cell reproduces itself with high fidelity, according to the ontologically correct (holistic) viewpoint. The ability to replicate DNA is unique to cells, not DNA. This understanding is critical for accurately portraying the life phenomenon (Figure 7.6).

Biological organization is hierarchical, and rational behavior is a property of the entire system, not its parts. Though the lifespan of an individual (the time from division to division) is short, the lineage is potentially (proper environment being provided) immortal. An asexual entity's long-term existence is thus an organism level lineage, a linear sequence of individual organisms.

Many give an import to the fact that asexual organisms can exchange genetic material now and then. However, the biological significance of such interactions is quite different. There is sexual reproduction but no sex. Asexual entities do not form biological species according to Mayr-Dobzhansky, a higher-ranking entity.

They don't need this complication because they can reproduce themselves with high fidelity while acting alone. If a mutated individual survives, it starts a new lineage that may compete with or swap with the previous lineage. The metaphysics of a biological species as an individual is a complex philosophical and epistemological problem with a long history of discussion.

The ontological aspect of the problem: species as a form of existence has been observed here. As the basic definition of species, it has been observed that, of Ernst Mayr: "Species are groups of interbreeding natural populations that are reproductively isolated from other such groups." Both traits "interbreeding" and "reproductively isolated" are obligatory.

Many people, including myself, consider biological species to be ontological individuals. Michael Ghiselin articulated this point of view clearly. A species, like an individual organism, goes through ontogenesis, which includes birth, infancy, adolescence, maturity, aging, and death. This eternal existence is carried out through a series of generations.

This sequence is commonly referred to as microevolution, which may be misleading because the essence of species ontogenesis is not evolution. In the event of success, a species' ontogenesis results in stasis (i.e., a halt in evolution) that can last tens of millions of years or more. Theoretically, the potential immortality cannot be excluded.

Generation of deme is a self-reproducing entity. It means that generation of deme, not the deme, is the ontological individual. So, a deme is equivalent to an asexual individual organism in its role in survival and creation of the lasting entity, which is a lineage, sequential row of self-reproducing entities, sequential raw of generations of deme.

Species is an organization. The bonding of intra-species components (individual organisms) is accomplished through behavior. By analogy with ionic, covalent, hydrogen, and other bonds, the term "behavioral bond" was proposed to describe the interaction between organisms. Behavioral bonds contribute to species cohesion.

Species-specific behavior implies the operation of special connections between individuals, which transform the species into a survival-oriented organization. These are the connections that are primarily responsible for interbreeding and reproductive isolation, which make a species a genetically closed monad. The mutual affinity of organisms determines reproductive isolation.

The affinity is not limited to choosing a mating partner; it includes all intraspecies interactions as well as interspecies interactions. Reproductive isolation and preventing inbreeding are two opposite "forces," analogous to the attraction and repulsion, the proper balance of which is a necessary precondition for a species existence.

Inbreeding and promiscuous sexual behavior are both harmful to a species. What exactly is "the proper balance"? What is the desired outcome of this balancing? It is the ideal species gene pool. A species as a whole is a closed genetic system. Parts of the species (groups, demes) may be capable of meaningful existence in nature.

This ability is analogous to plants' and lower animals' ability to regenerate the entire body from parts. As such, this is not a case of speciation. The lack of physical skin makes it difficult to understand a species as a whole. However, it is simply a matter of habit and imagination. Behavioral bonds are extremely common in biology.

They bring together families, groups, tribes, armies, businesses, states, and humanity. The transition from genome to genetic pool is far from complete. A genetic pool, like a genome, is a non-substantive entity. It is reproduced as a part, an attribute, by a generation of the deme. Individual organisms' biological status is drastically altered by the need to reproduce a genetic pool.

Asexual organisms are self-sufficient sovereign players on the life stage. A monad is what it is. It has the ability to reproduce itself through generations acting alone. A sexual individual is a law-obedient citizen of the multi-organismic realm. It must collaborate. This collaboration is not limited to finding a sexual partner and raising a child.

The final goal is transmitting an optimal gene pool. This final goal requires the submission of the entire species organization, including individual organism behavior. What is the best gene pool? The most general answer is as follows: the pool that ensures the next generation's survival. Of course, this is an overly broad response.

Survival is a function of the organization as a whole, not of a gene pool, i.e., of deme generation. The term "robustness" refers to a property that characterizes perfect organization. The ability to perform successful ontogenesis despite mutational and environmental perturbations is defined as robustness. Robustness resources are derived from all levels of biological organization.

At the species level, the diversity of organisms within the species is the most important factor in creating robustness. This factor has two effects on species survival. Immediately, the material for the creation of perfect species organization is the diversity of organisms. The same diversity is the prerequisite for creation of an abundant gene pool bestowing species with genotypic plasticity.

This type of diversity is an emergent trait of a species that must be selected. Not all differences are created equal. However, it is possible to argue that a lack of diversity would result in rapid extinction of the species because such homogeny negates all of the benefits of sexual reproduction.

The preceding discussion shows that Darwinian selection of individual organisms, the conquerors in intraspecies competition, does not work in the case of sexuality. There are numerous arguments against "selection of the best," and this is one of them: such selection would effectively eliminate diversity.

Manufacturing individual genomes from the continuously shuffled population gene pool rather than direct copying of the ancestor's genome is a critical feature of sexual reproduction. The primary advantage of this method of reproduction is self-evident. Though large genomes cannot be precisely replicated, it is always possible to create one errorless genome from two with errors.

Furthermore, the degeneracy of the link structure-function implies that functionally robust genomes may contain a variety of sequences. That is, the inherent property of sexual reproduction is the creation of a large diversity of genomes and thus individual organisms.

The genotypic plasticity–the ability to change the population gene pool configuration reversibly in different environments–is the next emergent species property. This is a significant factor in species stability. This is the most fundamental level of cooperation–genetic.

While the genome of a sexual organism cannot be replicated with sufficient accuracy, a genetic pool can be replicated with sufficient accuracy. All of the well-known complications and difficulties of sexual reproduction are justified by this ability to reproduce accurately, because only accurate reproduction provides longevity.

The "picking over" mechanism of genome reproduction promotes genome and individual diversity. It is evident, however, that this diversity inevitably includes a high proportion of genomes (and corresponding individuals) of low vitality doomed to perdition.

Unfortunately, this was interpreted in the classic Darwinism (and in Neo-Darwinism as well) in the spirit of Malthus's idea of exponential growth of populations leading to the competition for resources: struggle of everyone against everybody. Natural selection was founded on the premise that organisms produce more offspring than can survive.

As a result, organisms must compete with one another. This competition was interpreted as a driving force in biological evolution, leading to the continuous improvement of biological entities. It was emphasized that the fiercest struggle must occur between individuals of the same species because they have the same needs.

This erroneous interpretation gave rise to the most heinous forms of social Darwinism, as well as an anti-evolutionary attitude among many intellectual and spiritual leaders. In reality, the apparent "extra" progeny is a compensation for the poor fidelity of genome reproduction and the organisms' random death.

Random death, i.e., death caused by neither competition nor genetic defects, may be widespread in the reproduction strategy. Individual organisms' differential survival in both cases is not due to resource competition. First and foremost, they live or die based on their own merits, regardless of the presence of other people. They are not usually killed or starved to death by their peers or rivals. The biotic entities' stability is determined not only by their physical durability, but also by their expedient behavior. They are organizations with a survival function. The Universe evolves through the interaction and cooperation of entities, which gives rise to its complexity and hierarchical structure. The major transitions in biological evolution (prokaryotic cell → eukaryotic cell → multicellular organism → biological species) are the steps of cooperation.

Though a complex entity is made up of simpler entities, it is not simply an aggregate of the included entities. It is a qualitatively new form of existence; it is a higher-ranking organization. In biology, hierarchy does not only refer to complexity or heterogeneity. It implies that their parts are functionally predestined for the sake of the whole.

Survival of the parts is critical to the survival of the whole. As a result, constituent entities should only be included in higher entities in an appropriately transformed configuration. The higher-ranking organization's operating principles are not always related to or derivable from the properties of the parts or their internal operating principles.

That is the principles organizing an upper rank are novelties. They are not necessarily predictable from the rank below. The organizing constraints of living entities, on the other hand, emerged as a frozen chance and cannot be deduced from any general principle or law.

They can only be understood in retrospect, in the context of their past. According to the preceding statements, the evolution of a higher entity cannot be adequately represented as self-sufficiency evolution of its constituents. The vector of selection for the constituent entities is the overall prosperity.

7.8. ALTRUISM

Altruistic behavior of an individual organism as defined by Neo-Darwinism is a behavior that diminishes its own fitness and enhances the fitness of other individuals. In its turn, fitness is defined as a relative fecundity of the individual. According to the same paradigm, "Evolution is based on a fierce competition between individuals and should therefore reward only selfish behavior. Every gene, every cell, and every organism should be designed to promote its own evolutionary success at the expense of its competitors."

If this is the case, altruism must be impossible. However, altruism is widespread. Another inconsistency is that sexual organisms are unable to reproduce. To cope with this snag, G. C. Williams and then R. Dawkins announced the gene as a unit of selection. The gene appeared to be a perfect immortal replicator. The kin selection and reciprocal altruism hypotheses provided a formal explanation for how evolution could favor altruism despite fierce competition between individuals. Of course, it did not prove that an individual organism or gene can truly serve as a unit of selection.

Most biologists, being not trained in mathematics, had been pressed to take for granted this logical trick under the label of "gene centered view." Unfortunately, it achieved worldwide success among population geneticists and molecular biologists, burying the holistic understanding of life and evolution for dozens of years.

Meanwhile, an altruistic behavior, which is ubiquitous among people and other animals, keeps being a headache for the evolutionary biologists. The Hamiltonian pill has helped to alleviate the headache, but the phenomenon remains enigmatic and fuels endless and mostly fruitless debates.

It's difficult not to notice the various forms of cooperation, mutual aid, friendship, and love at every turn. For this, one must specially train his/her imagination in the reductionistic logics or reject the phenomenon ironically as did Michael Ghiselin: "Scratch an 'altruist,' and watch a 'hypocrite' bleed."

Meanwhile, the apparent problem of the origin and maintenance of altruism is the result of the gene-centered point of view and reductionistic philosophy. The reductionistic methodology is insufficient for working with the hierarchically organized world of life. The idea that gene is a replicator causes bewilderment.

Figure 7.7. Image showing Altruism–Gorilla and child.

Source: Image by flickr.

It is not a living entity or even a self-replicator. Gene is replicated. It is a replica or a template. Much like all the other cell constituents: RNAs, polypeptides, organelles, genes are manufactured by cells as well. Only a self-reproducing substantive entity can serve to be a unit of selection (Figure 7.7).

Course in favor of deme (or group) selection as the only meaningful level of selection for obligatory sexual organisms. Of all the entities deme is the lowest substantive entity that reproduces itself with a high fidelity. Opponents of the group selection reject it as a greatly slow process.

7.8.1. A Successful Species Ceases to Evolve

Higher-level selection was easily beaten by lower-level selection. First, the group selection process may be quick enough: a generation of population (the unit of reproduction) is of the same longevity as an individual organism; second (and uppermost), the lower-level selection in itself is the destructive side of the overall process.

If it is not trumped by the higher-level selection, the group simply will not go through. The mechanism of species evolution (microevolution or species ontogenesis) presumes a group selection. Successful groups may thrive, while less successful groups may contract or become extinct.

Again, competition and struggle between groups may play little role in their fate: they parish or prosper primarily on their own merits, e.g., because of the constituent entities' selfish or altruistic behavior. For a comprehensive discussion of the group selection problem. The concept of multilevel selection has now gained widespread acceptance.

It is a positive step forward. There is, however, some inconsistency. Given that the higher level of selection operates, lower-level selection must be prohibited because it can result in nothing but casualties such as parasitic DNA or a malignant cell. The so-called ultra-selfish genes are parasites that have a negative impact on the host.

They, along with other parasites and harmful mutations, are manifestations of nature's destructive force. In action, evolution is a never-ending battle against this force. And cooperation is the most fruitful way to fight this battle. Biological species are organizations formed by the cooperation of individual organisms.

A hierarchical organization presumes submitting behavior of parts in favor of the whole. Selection presumes the selection of genes but the vector of selection is "for the good of species," not "for the good of gene"! Because a gene is not a living entity, it is not an organization with a survival function!

It's strange to insist on such an obvious statement. These two items are related. If they not coincide, the gene will be rejected "for the good of species." The opposite (rejection of species for the good of gene) is nonsense.

7.9. CONCLUSION

In this chapter, two types of existence have been discussed - animate and inanimate. No existence lasts forever. The second law of thermodynamics predicts that everything in the Universe will eventually dissipate. Because the Universe's components interact with one another, it does not dissipate instantly. The Universe as a whole is constantly changing.

The four fundamental physical interactions generate numerous entities, delaying the Universe's immediate dissipation. The lengths of various objects in the Universe range from infinitesimal fractions of a second to billions of years. Through the course of the evolution, the Ephemeral forms are replaced by the more permanent ones.

This principle–survival of the fittest–appears to be a tautology, but it is the great tautology: everything genuinely new emerges from it. Longevity is a quantitative assessment of existence. Inanimate entities are the most likely configurations of matter in a given situation; their physical durability ensures their longevity.

Organisms are low-probability matter configurations; they are physically fragile, extremely complex, low-entropy systems. Their existence necessitates a unique explanation. We are accustomed to believing that the entire Universe is comprised of two interconvertible essences: energy and substance.

Some people have recently begun to discuss the third fundamental essence–organic information, which is neither energy nor matter. The term "information" may not be the most suitable one. It may be confused with the homonym used in the information theory.

Shannon's information is devoid of meaning whereas the meaning is just what we are interested in context of biology. An adequate term for the third fundamental essence would be "mind." The living world resists chaos through rational behavior. It has been observed that behavior is sensible when the behaving entity's survival is at stake. Inanimate evolution could be portrayed as Nature's self-construction: matter from energy; biological evolution could be portrayed as Nature's self-knowledge: mind from matter. Physics' autocracy ends at the boundary between the inanimate and the biosphere, where the world of sense and knowledge begins. All living entities are organizations that exist solely for the purpose of survival.

Self-reproduction ensures the longevity of a living entity. The sensible behavior is based on knowledge. The entire knowledge a living entity enjoys is the knowledge of how to reproduce itself. The self-reproducing entity in asexual organisms is an individual organism. A prokaryotic cell is the most basic living entity.

Eukaryotic cells, multicellular organisms, and biological species are examples of more complex living entities. Asexual and sexual reproduction are the two types of reproduction. Asexual organisms (typically prokaryotic cells) reproduce themselves with a high fidelity sufficient for the lineage's potential immortality.

Higher organisms are far too complex to reliably reproduce on their own. They work together to form a higher-ranking multi-organismic entity–a biological species. An organism is a substantive entity in the case of asexuality, whereas a sexual individual is an attributive entity that exists as a part of the higher rank entity–deme. The deme generation is self-replicating.

The formation of individual organisms' genomes by randomly selecting them from the continuously shuffled gene pool rather than direct replication of the ancestor's genome is a critical feature of sexual reproduction. This process invariably results in individual organisms with varying survival abilities.

In general, they live or die based on their own merits, regardless of the presence of other entities. This is a period of purification or intrinsic selection. A species' evolutionary success is determined by the perfection of species organization, which includes cooperative interaction between individual organisms. This cooperation is an example of the fifth life-specific fundamental interaction.

REFERENCES

1. Avise, J., (2014). 1935 biological species. *Conceptual Breakthroughs in Evolutionary Genetics*, 43, 44. [online] Available at: https://www.sciencedirect.com/science/article/pii/B9780124201668000208?via%3Dihub (accessed on 27 June 2022).
2. Prokhorovich, V., (2013). Biological species as a form of existence, the higher form. *The Species Problem-Ongoing Issues*. [online] Available at: https://www.intechopen.com/chapters/42176 (accessed on 27 June 2022).
3. Shcherbakov, V., (2010). Biological species is the only possible form of existence for higher organisms: The evolutionary meaning of sexual reproduction. *Biology Direct, 5*(1), 14. [online] Available at: https://biologydirect.biomedcentral.com/articles/10.1186/1745-6150-5-14 (accessed on 27 June 2022).
4. Shcherbakov, V., (2013). *Biological Species as a Form of Existence, the Higher Form*. [e-Book] INTECH. Available at: https://www.researchgate.net/publication/256191859_Biological_Species_as_a_Form_of_Existence_the_Higher_Form (accessed on 27 June 2022).
5. Wool, D., Paz, N., & Friedman, L., (2020). Ernst Mayr: The biological species and speciation. *Milestones in the Evolving Theory of Evolution*, 278–283. [online] Available at: https://www.taylorfrancis.com/chapters/mono/10.1201/9781003023869-36/ernst-mayr-biological-species-speciation-david-wool-naomi-paz-leonid-friedman (accessed on 27 June 2022).

CHAPTER 8

Ethical Implications of Sociobiology

CONTENTS

Ethics is a universal human concept. Humans have moral values, which means they accept norms that determine whether their actions are right or bad, good or evil. The specific rules by which moral behaviors are assessed differ to some extent between individuals and civilizations (although basic norms, such as not killing, not stealing, and honoring one's parents, are ubiquitous and possibly universal), but value judgments are passed in all societies.

This universality begs the question of whether moral sensibility is a component of human nature, another facet of our biological make-up; and whether ethical principles are the result of biological evolution instead of just being handed down via religious and cultural traditions.

The purpose of this chapter is, first and foremost, to broaden and improve our knowledge of human moral functioning by investigating contemporary sociobiological and neuroscientific thinking about human conduct. Secondly, it addresses sociobiology's implications for moral teaching. To accomplish these objectives, our research begins with a critical examination of the sociobiological assumption that social behavior may be handled physiologically.

Several research is also presented that aim to integrate sociobiology and other closely related sciences, such as neurobiology and evolutionary psychology, with moral psychology. Finally, we seek to illustrate the potential use and applicability of sociobiology in the fields of moral psychology and moral education in this work.

8.1. INTRODUCTION

Aristotle and other classical Greek and Roman thinkers, as well as Thomas Aquinas and the scholastics, believed that we are ethical beings by nature. Man is both homo sapiens and homo moralis. However, biological evolution introduces an essential diachronic dimension. Animals do not exhibit ethical conduct (at least not to all animals and not to the same extent as to humans).

Even if we agree with Aristotle and Aquinas, the following issues remain: When did the ability to act ethically emerge? And why did it change? Is it a natural result of other traits (such as intellect) or was it actively pushed as a direct target of natural selection? Natural selection theory is applied to behavior in sociobiology.

It states that an animal's behavior, like its structure, is the result of biological evolution by natural selection. Any behavioral phenotype arises from the combination of genetics and environmental factors (which include other members of the same species and, in the case of man, his material and symbolic culture).

Culture is a whole new evolutionary ballgame for humans; cultural evolution is significantly faster than Darwinian, genetic evolution. Human culture, on the other hand, does not exist independently from biological evolution; it arose out of it and is closely linked to it.

One might believe that, after the Scopes trial in 1925, the aforementioned claims are no longer contentious. However, sociobiology has been challenged as a poisonous, racist, and reactionary philosophy, much like Darwinists were accused of being communists, anarchists, and revolutionaries during the Scopes trial.

The New Left of "Science for the People" has joined forces with backwoods fundamentalism in rejecting natural selection as a factor in the evolution of human social behavior. No sociobiologist I know rejects the role of culture in humans—or "tradition" in many higher vertebrates—but many individuals, especially social scientists, continue to dismiss Darwinian evolution's relevance to human social behavior.

What is at risk is not man's individuality. Every species is distinctive in some way; otherwise, it would not be a distinct species. It is also undeniable that humans have evolved in ways that distinguish us from other animals, such as symbolic language, rational decisions, conspiratorial conduct, productive, and destructive technology, and environmental control.

What sociobiologist fail to accept is the widely held belief among social scientists that human behavior is infinitely malleable and unconstrained by genetics. What, if any, ethical consequences do sociobiology have? A sociobiological perspective of human behavior does not logically lead to any ethical implications. Sociobiology is not a philosophical system.

It is devoid of teleology. It makes no assumptions about whether evolution—or survival, or reproduction, or anything else—is beneficial or even serves any purpose. Sociobiology and the natural selection theory that underpins it can help explain why creatures evolve over time. Evolution has been a fundamental principle of biology since Darwin's discovery of natural selection.

Furthermore, it has had an impact on many domains other than biology, including other natural sciences, social science, and the humanities. For example, his concept has been transferred into politics, prompting some to distinguish between the Darwinian left and the conventional left (Singer, 1999). Indeed, Darwin's theory has begun to pervade scientific discourse on human conduct (Figure 8.1).

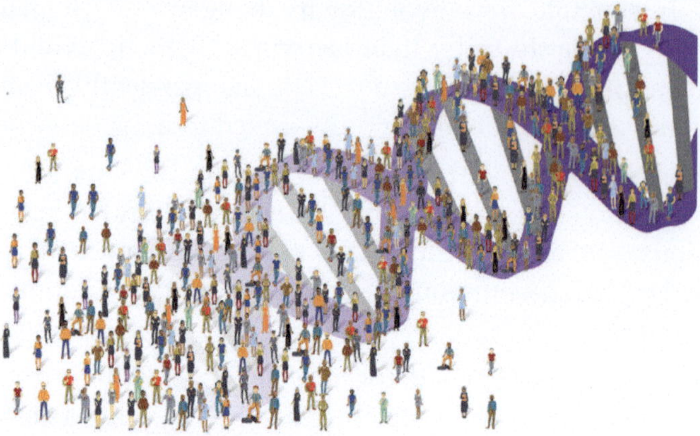

Figure 8.1. Centers of excellence in ethical, legal, and social implications research (CEER).

Source: Image by flickr.

Contemporary biological scientists interested in evolutionary explanations for human behavior have stressed that effective humanities or social sciences studies begin with an appreciation of the human condition. Wilson (1978), for example, stated that without evolution, the humanities and social sciences are restricted descriptions of surface phenomena, just as astronomy, biology, and mathematics are without algebra.

Thus, an increasing number of studies deal with scientific aspects of the human brain and moral functioning, reflecting the growing scholarly tendency to approach human nature scientifically and attempt to develop a natural science-based understanding of the human mind, as is happening in neurobiology and neuroimaging.

8.2. SOCIO-BIOLOGY AND THE ORIGINS OF ETHICS

The sociobiological theory is concerned with "social" animal species, which establish groups, develop a division of labor, and collaborate in order to survive in the battle for survival. As a result, these animals appear from the outset to be equipped with moral characteristics.

"Social" animals have qualities comparable to humans not only because they are born and die, develop, reproduce, and become ill, nor because they interact with other beings of the same and even other species, but because, as they perform all of these various functions in their own unique ways, they exhibit emotional and behavioral patterns that are strikingly similar to those of humans.

The Darwinist theory is a major source of sociobiology which, after opposing creationism and upholding the idea of species continuity, rejects the existence of broad qualitative distinctions between humans and higher animals. Therefore, Darwinism is rationally forced to admit that morality arises from evolutionary mechanisms shared by all living creatures.

This theory's contribution to the problem of the origins of ethics comes, first and foremost, in the search for an explanation through a scientific process of analysis, a method that naturalizes the moral standing of species. Species must earn their right to survival through everyday conflict, according to Darwinism (Figure 8.2).

Figure 8.2. Image showing ethics.

Source: Image by pixabay.

No species is innately more valuable than any other; natural selection will demonstrate the worth of each species and, if successful, will manifest itself in their perpetuation over time. The only criterion by which morals may emerge, according to Darwinist theory, is reproduction.

Wherein each species defines its "tactics" of battle in order to ensure survival in this way, a final moral valuation would consider these requirements. Sociobiology, as an extension of traditional Darwinism (Ruse, 1987), strengthens this contribution.

8.2.1. Continuity and Ethics in Contemporary Socio-Biology

Sociobiologists recognize the presence of a continuum in evolution as a universal process, one of the results of which is the creation of ethics and moral conduct. M. Ruse likewise adheres to this viewpoint when he says, "the criteria of ethical behavior are determined by the causal rules of evolution," and when he says that the evolutionary process gives moral guidance given that there is intrinsic worth in the current condition of the biological world.

Because these traits are a product of social interactions as well as human evolution, and because there is no distinction between human evolution and the rest of biological evolution, social growth is equated with moral advancement and both with biological progress.

It has also been argued that the Darwinist concept of animal-human continuity is ant essentialist since it does not see humans as being so unique that they have an essence that is significantly distinct from the rest of the living species. There is some ambiguity in this situation. A qualitative difference between two entities does not entail that they are diametrically opposed.

Entity B emerges as a result of A's development following a qualitative shift, and it can keep and develop many of A's traits. What distinguishes B qualitatively from A is the prevalence in B of features never seen previously in A. It requires B's behavior to be observed using rules and techniques distinct from A's.

This is what distinguishes a qualitative distinction. In terms of the evolutionary roots of morality, the attribute of suffering and having emotions in many creatures does not indicate that all of them have a moral sense, even in a basic form.

The moral sense will emerge in humans as a result of feelings, emotions, and social activities, but it cannot be regarded just as a function of these features in their animal manifestation. Human moral sense gains independence and a distinct development.

8.3. SOCIOBIOLOGICAL ETHICS AND SURVIVAL

It was examined, that committing the naturalist error in sociobiology and its translation into ethical relativism are handled from the outset by the survival principle, which gives rise to morality predicated on the idea of "the ends justify the means."

One of its main sociobiological theses, in which a specific concept about the functioning of human society is projected onto the entire living world, asserts that organisms must always struggle against a hostile environment in order to survive and leave offspring, thereby generating individual interests and conflicts.

All through history, nature has seemed like a malevolent element beyond human control. When people attempt to overcome this problem, they adapt to what nature demands of them; nevertheless, in doing so, people also adjust nature to their requirements, dominate it, and are capable of generating new conflict situations that are not the result of nature's permanent condition but of human activity on it.

Mankind is capable of influencing their environs; they are not just subject to nature. Thus, I feel it is more exact to characterize the condition of living creatures as a continuous and complicated process of interaction that might cause more or less extensive oscillations of hostility/benefit and adaptation/inadaptation.

These oscillations result in oscillatory behaviors that Darwinists label as selfishness, altruism, competition, and cooperation. These oscillations occur at numerous levels, ranging from the individual to the interpopulation. A rigorous view of living beings as either extremely selfish or extreme altruists adds to a deficient and reductionist understanding of the human world and therefore should be rejected.

Depending on the circumstances, an individual may act selfishly on one occasion and altruistically on another, spanning all intermediate points between the two extremes throughout time. The answers to all of these circumstances are quite complicated; to evaluate each one, all of the aspects and factors that intervene must be integrated.

Sociobiological analyzes on the origins of ethical behavior have ignored the profound dynamic of human relationships and their constant transformations, because it begins with a prior conviction of a constant in the human being that, as a result, remains unchanged in his relationships with nature, and because it treats the human being abstractly, without drawing society or discriminating its components in space and time (Figure 8.3).

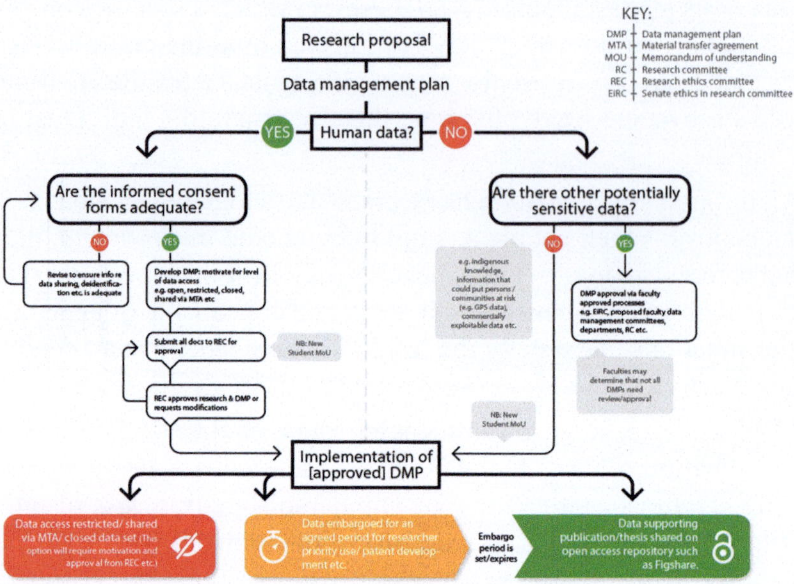

Figure 8.3. UCT data management planning (DMP) for ethics approval process.

Source: Image by Wikimedia commons.

8.4. MORAL EVALUATIONS AND MORAL NORMS

The topic of whether ethical conduct is genetically determined might apply to one of two issues:

- Is the aptitude for ethics–the inclination to assess human activities as either moral or bad–dictated by human biology?
- Are the ethical rules or systems that humans accept biologically determined?

The first point is more basic; it asks if man's biological constitution is such that individuals are compelled to form moral judgments and embrace ethical ideals, to label certain activities as right or immoral. Affirmative responses to the first question do not always indicate the answer to the second question. Regardless of whether humans are inevitably ethical, it remains to be discovered if certain moral precepts are in fact defined by man's biological nature, or whether they are selected by society or by individuals Even if we decide that humans cannot escape possessing moral standards of behavior, the selection of the specific criteria employed for judgment may be arbitrary.

> The need for moral values does not necessarily inform what those values should be, just as the ability to communicate doesn't really dictate the language one will speak. The theory that shall be present is whether humans are ethical beings by nature; that human assess their actions as good or wrong, moral or immoral, as a result of their exceptional intellectual skills, which include self-awareness and abstract reasoning.

These intellectual abilities are evolutionary byproducts, yet they are uniquely human. As a result, It will argue that ethical conduct is not causally tied to animal social behavior, including kinship and reciprocal "altruism."

The second thesis that will be advanced is that the moral standards that we use to judge whether certain behaviors are morally acceptable or morally terrible (as well as the justifications that might be used to explain the moral norms) are the result of cultural development, not biological evolution.

Moral norms fall into the same group of phenomena as political and religious organizations, as well as the arts, sciences, and technology. Moral rules, like other artifacts of human civilization, are frequently congruent with the innate predispositions of animals and humans.

Figure 8.4. Ethics word cloud.

Source: Image by pixabay.

This congruence between ethical rules and biological inclinations, however, is neither required nor universal: it does not apply to all ethical norms in a particular community, fewer all human communities. Moral codes, like all cultural systems, rely on the presence of human biological nature and must be consistent with it in the sense that they cannot be opposed without supporting their own destruction.

Furthermore, when moral rules are congruent with physiologically conditioned human actions, they are easier to accept and maintain. However, moral norms are independent of such actions in the sense that some standards may not support and may even harm the survival and reproduction of the person and its genes, whereby survival and reproduction are the goals of biological evolution (Figure 8.4).Discrepancies between accepted moral principles and biological survival, on the other hand, must be strictly limited or lead to the extinction of the communities adopting such inconsistent norms.

8.5. THREE NECESSARY AND SUFFICIENT CONDITIONS FOR ETHICAL BEHAVIOR

The issue of whether our biological makeup influences our ethical behavior must be addressed in this yes. By "ethical conduct," I mean the desire to categories human behaviors as either good or bad, rather than good or bad behavior (i.e., choosing to do what is perceived as good instead of what is perceived as evil).

Humans display ethical conduct by nature because their biological makeup defines the presence of the three necessary and mutually sufficient criteria for ethical action. These are the following conditions:

- The capacity to predict the effects of one's own actions;
- The ability to form value judgments; and
- The ability to choose between various courses of action. It was to investigate each of these talents and demonstrate that they exist as a group as a result of humans' exceptional intellectual capacity.

The ability to predict the effects of one's own actions is the most basic of the three requirements for ethical behavior. Only if expect that pushing the trigger will result in the bullet striking and killing an adversary can the action of pulling the trigger be considered evil. Pulling the trigger is not a moral action in and of itself; it becomes such because of the consequences.

Only if Someone anticipates these effects does my behavior have an ethical dimension. The capacity to predict the repercussions of one's actions is directly tied to the ability to create the link between means and purposes; that is, to recognize a mean precisely as a mean, as something that serves a certain objective or purpose.

This ability to connect means and ends necessitates the ability to foresee the future and build mental representations of realities that are not visible or do not yet exist. The ability to connect means and objectives happens to be the fundamental intellectual talent that has enabled the advancement of human culture and technology.

The evolutionary origins of this ability may be discovered in the emergence of the upright position, which changed our ancestors' anterior limbs from locomotion organs to manipulation organs. As a result, the hands evolved into organs capable of building and using things for hunting and other activities that benefited survival and reproduction, i.e., raised the reproductive fitness of their carriers.

The production of tools is dependent not only on manual skill but also on accurately understanding them as tools, as items that aid in the performance of certain tasks, that is, as means to a goal or purpose: a knife for cutting, an arrow for hunting, an animal hide for protection against the cold.

Natural selection favored our bipedal ancestors' intellectual ability since higher intelligence allowed the perception of tools as tools, and hence their fabrication and use, resulting in improved biological survival and reproduction.

Our predecessors' intellectual talents evolved over three million years or more, steadily improving the capacity to connect means with their objectives and, hence, the capability of building ever more complicated instruments serving distant purposes.

The capacity to predict the future, which is required for ethical behavior, is therefore strongly linked to the development of the ability to make tools, a skill that has generated the advanced technology of contemporary society and is primarily responsible for mankind's success as a biological species.

Humanity has expanded throughout the whole globe, except for the freezing wastelands of Antarctica, and has become the most abundant species of mammal. Numbers may not be a perfect blessing, but they are, nevertheless, a measure of biological success. The ability to form value judgments, to regard certain items or behaviors as more acceptable than others, is the second criterion for the existence of ethical conduct.

Only if consider an enemy's death to be preferable to its survival (or vice versa) can Then consider the action that leads to death to be moral. If the alternative outcomes of an action are value neutral, the activity cannot be described as moral. The ability to make value judgments is dependent on its ability to abstract, or to perceive acts or objects as representatives of generic classes.

This makes it possible to compare items or behaviors and determine which are more appealing than others. The ability to abstract necessitates sophisticated intelligence, which appears to exist only in humans, and seemingly only in them. The ability to select between various courses of action is the third requirement for ethical behavior.

Pulling the trigger is only moral if there is an option not to pull it. An essential action outside our control is not a moral action: blood circulation or food digesting are not moral activities. The existence of free will is a hotly debated topic among philosophers, and this is not the place to go into the arguments. I will simply make two points that are common sense proof of the existence of a free will.

One is our own experience, which shows that having the option to pick between possibilities is actual, not just apparent. The second point to examine is that when we are confronted with a circumstance that needs us to act, we are psychologically capable of exploring different courses of action, so expanding the space within which we can exercise our free will.

In any event, there would be no ethical action if there was no free will; morality would be fiction. The argument to make here is that free will is contingent on the existence of a well-developed intelligence, which allows one to consider multiple courses of action and select one or the other based on the expected outcomes repercussions.

In conclusion, ethical conduct is a feature of human biology and hence a consequence of biological evolution. However, I find no evidence that ethical conduct evolved as a result of being adaptive in and of itself. But don't understand how judging particular behaviors as good or evil (rather than just preferring certain actions over others or rating them based on their practical effects) would improve the assessors' reproductive fitness.

And also, don't understand how there could be "incipient" ethical conduct that would be encouraged by natural selection. The three prerequisites for ethical behavior are expressions of superior intellectual talents. It appears that the aim of natural selection was the growth of these superior intellectual abilities. Natural selection favored this because the development and usage of tools boosted our biped ancestors' strategic position. When bipedalism developed and tool use and tool-making were available, people who were more effective in these functions had a higher chance of biological success. The biological advantage offered by tool invention and usage remained long enough for intellectual capacities to grow, eventually providing the exceptional development of intelligence that is distinctive of Homo sapiens.

8.6. THE EVOLUTION OF INFORMATION PROCESSING AND THE QUESTION OF ANIMAL ETHICS

Human intellectual talents may be viewed as one endpoint of an evolutionary continuous and progressive process. A steady rise in the ability to collect and absorb information about the external world is an evolutionary tendency that is notably visible in animal lineages. This skill is adaptive because it permits the organism to respond to environmental changes in a flexible manner (Ayala, 1982a).

Certain single-celled creatures have a primitive capacity to acquire and absorb information about their surroundings. A paramecium moves in a sinuous course, consuming the bacteria it finds. When it comes into contact with adverse circumstances, such as excessive acidity or saline in the water, the paramecium checks its progress, rotates, and begins moving in a new direction.

This is a strictly negative reaction: the paramecium does not seek food or a pleasant habitat, but rather avoids undesirable ones. The single-celled alga Euglena possesses a sensitive area that allows it to orient itself towards the direction of light, giving it a stronger capacity to interpret environmental information (Figure 8.5).

Figure 8.5. Brexit–animal welfare.

Source: Image by flickr.

Euglena moves in a directed manner; it not only avoids undesirable settings but actively seeks them out. An amoeba shows continuing development in the same direction; it moves protected from sunlight and actively chases particles of food. The ability to collect and absorb environmental information has not risen with time in all evolutionary lineages.

Bacteria today are no more sophisticated in this regard than their ancestors one billion years ago. Many evolutionary lineages made minimal development in the early stages but made no more progress throughout their lifetimes. Animals are more advanced than plants in general; vertebrates are more advanced than invertebrates; mammals are more advanced than reptiles, which are more advanced than fish.

Vertebrates can collect and interpret significantly more complex signals and create a much wider range of responses than invertebrates such as insects and other arthropods. The capacity of animals in general to collect and digest information about their surroundings is founded in the nervous system and the brain, which integrates the sensory impulses provided by the nerves and organizes the necessary reactions.

The vertebrate brain has a massive number of associative neurons in a very sophisticated architecture. Progress in the ability to collect and process environmental information is associated with a growth in the size of the cerebral hemispheres as well as the emergence and development of the neopallium in vertebrates.

The neopallium is an organ which connects and coordinates impulses from all receptors and brain areas. The neopallium originally appeared in reptiles. In animals, it has grown to become the cerebral cortex, which encompasses the majority of the cerebral hemispheres (Figure 8.6).

Figure 8.6. Animal welfare.

Source: Image by maxpixel.com.

Because vertebrates have bigger brains than invertebrates, they can have a greater number of neurons dedicated to information storage or memory. The relative size and absolute complexity of the brain, particularly the cerebral cortex, reaches a peak in humans, who have a far better capacity than any other organism to perceive and integrate information coordination and flexibility in response to what is seen.

The unusual growth of the brain has equipped humans with intellectual abilities that enable abstraction and self-awareness, i.e., the objectivation of the thinking subject, or an individual's ability to recognize them self as an object.

8.7. MORAL NORMS: RELIGIOUS AND EVOLUTIONARY PROPOSALS

Ethical behavior is founded on man's basic nature also hypothesized that ethical conduct evolved as an indirect result of the evolution of exceptional intellectual capacities, rather than as a result of self-adaptation. Now comes the second question: if biological nature also decides the moral standards or ethical rules humans must follow (Figure 8.7).

Figure 8.7. Say no to animal cruelty.

Source: Image by pixabay.

The moral rules that humans use to determine whether specific conduct is right or wrong are determined by cultural evolution rather than biological evolution. These moral judgments are based on religion as well as other social customs.

8.8. SOCIOBIOLOGICAL ETHICS AS AN IDEOLOGY

Socio-biology doesn't quite address, or ever even defend, the issue of freedom. This never suggests examining the scope of free human activity, just its restrictions. The selfish and competitive person in a certain context is equated with humanity in general in this view, and hence the foundation of a very concrete moral code is identified with the foundations of all moral codes.

This is an example of an "ideological error," which occurs when a component is equated with the total and stripped of its historical context. The issue with sociobiological ethics is that it does not emphasize the material roots of morality and ethics; rather, it merely transfers to animals' particular human values specific to a historical moment and then proceeds to naturalize them both in humans and animals.

Such that, it adopts a specific worldview and regards itself as the only one that exists. If an evolutionist theory doesn't include history in its framework, it cannot answer the question of value continuity and discontinuity as dynamic phenomena. This produces a false awareness of the world, i.e., an ideology, and its integrating goals change into an incomplete, fragmentary, and decontextualized vision of the world.

Reproductive success is merely the definition of adaptation, strictly speaking, natural selection is not a theory but a tautology. In both their morphology and their behavior, thus, natural selection is the simplest and most general description of how living things change.

This only superficially understood when the specific mechanisms through which this happens (recombination through chromosomal crossovers, sexual selection and reproduction, meiosis, mutation,). In ascribing an intrinsic morality, ideology, or teleology to sociobiology than to astronomy or biochemistry one is no more justified, in any case.

As a strain of Social Darwinism did in the late 19th century, the critics of sociobiology come principally from self-styled leftist circles who fear that sociobiology will become an ideology of racism and conservatism. The specter of eugenics and genetic engineering feared by others.

It is possible, for instance, in behavioral traits between human groups, that significant genetic differences may be found and, it is highly probable that they may serve to justify many human prejudices, if found.

Because of genetic and biochemical differences, has spawned a multitude of ideologies and moralities, the fact that males and females of our species, and of other sexually reproducing species as well, behave differently, in part. A liberating, liberalizing, even revolutionizing influenced by sociobiology. Humans are beneath their cultural differences, by stressing how fundamentally alike, for instance, to racism sociobiology could be a powerful antidote. We could engineer the reduction of sexual dimorphism in humans and strike a blow for androgyny, or by using our knowledge of the causes of sex differences.

The possibilities are endless. We are especially sensitive to critics from the left because of our recent experiences with racism and genocide, but as a threateningly radical ideology in the 19th century religious fundamentalists saw Darwinism. To the content of scientific theories, our use of knowledge bears only accidental resemblance.

Knowledge is commonly used for self-serving purposes and that is perhaps the safest conclusion, it generally serves conservative ends, since it is often most effectively used by those in power. It is true that the scientists have ideological biases like the rest of mankind whether of the left or the right, and held values inevitably intrude on these privately, and bias, scientific inquiry.

The borderline between would-be scientific theory and ideology is frequently fuzzy it is also true that in social science, and therefore in scientific inquiry, the practice of questioning one's values and motives is a sound corrective to the intrusion of values.

As academics in other disciplines, sociobiologists cover approximately the same political spectrum that is clearly left of center on the American political scene with a center of gravity. To identify with sociobiology are entirely outside the current of sociobiological thinking, by some of the people whom critics of sociobiology have sought.

Arthur Jensen, for example, makes meaningless statements (e.g., to heredity rather than the environment, on proportions of IQ variance attributable) to which few if any population geneticists or sociobiologists, would subscribe aware of the complexity of the relationship between phenotype and genotype in intelligence.

Is some of our moral and ethical precepts there not be a sociobiological basis? Within broad limits, moral injunctions are congruent with evolutionary strategies of fitness maximization, some evidence suggests that. In a wide variety of cultures consider the double standard of sexual morality found.

To the extent that females of practically all sexually reproducing species produce far bigger and fewer to put it briefly, in the choice of mating partners and to appreciate quality rather than quantity of offspring, therefore more valuable, gametes than males, they can be expected to be more selective than males. Where the number of offspring is limited, where gestation is long, and where lactation further increases maternal, as opposed to paternal, investment in the young, the argument applies even more to mammals. It is not only human females who left holding the babies and play coy. In the last couple of decades seems to be moving toward a sexual morality that rejects the double standard our own society.

Our behavior in this respect is so rigidly programmed that our culture cannot modify it, the argument is clearly not that we have a gene for the double standard. There is a cultural codification of differential reproductive strategies of males and females in humans rather, I suggest that the double standard of sexual morality.

Moral standards are unique to humans so far as we know, but to sexually reproducing species, including our own, differential parental investment of males and females in offspring is general. Dissociates sexual and reproductive behavior in effect our recent technology of contraception. The risk of conception, is reduced dramatically for females, and, behold, sexual morality changes, even with an unfit partner.

8.9. CULTURE AND GENES

It is suggested by an example, that the linkages between culture and genes are anything but simple or mechanically deterministic. To alter drastically the course of our evolution, culturally, and even genetically we certainly have the capability. This is not to say that our behavior ceases completely to be biologically predisposed.

On the physiology of sexual arousal, it will be interesting to see whether the technology of contraception will have a feedback effect. The new culturally created conditions should over a few' generations reduce sexual differences in speed of arousal, assuming that slower female arousal was an adaptive response to greater female cost of reproduction., Genes not only act on culture but culture acts on genes.

It works both ways. A possible sociobiological basis for ethics is far broader in scope in a second illustrations. It concerns the complex set of social norms that is present in varying degrees in all human societies and underlies social existence itself. Beyond the nuclear family to kin groups almost invariably morality extends (lineages, clans) and to still larger groups (racial, ethnic, linguistic, religious, national), at a minimum, moral rules enjoying us to honor our father and mother and cherish our children (Figure 8.8).

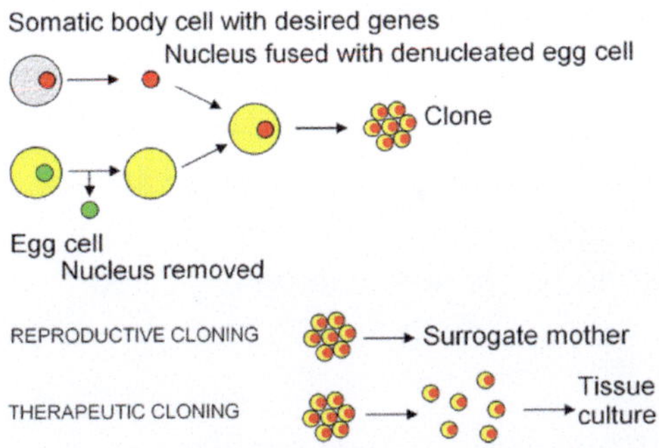

Figure 8.8. Reproductive and therapeutic cloning.

Source: Image by Wikimedia commons.

At least in theory, a few ethical systems have encompassed, our entire species: Hinduism, Buddhism, and Jainism to all living things, go so far as to extend moral precepts. Generally, however, the less well they work in practice, the more sweeping the scope of moral precepts.

It frequently uses the idiom of kinship, when morality is extended beyond the nuclear or extended family: the emperor is the "father" of his subjects, members of a race or of a religious sect are "brothers" and "sisters" to each other; we allow ourselves to be slaughtered in time of war; and so, on it is for the "fatherland" or the "mother country."

8.10. ENLIGHTENED SELF-INTEREST

They parallel in their effects what sociobiologists call "kin selection," "inclusive fitness," and, rather misleadingly, "altruism," insofar as these moralities are effective in eliciting "altruistic" behavior-beneficial to "alter" at some cost to "ego." Because it refers in fact to the ultimate form of genetic selfishness altruism is a misnomer.

We share with the beneficiaries of our altruism Kin-selection theory says in effect that our altruism is proportional to the number of genes. We indirectly, and to the extent that we are related, enhance our own fitness or, more precisely, that of our genes, by increasing the reproductive fitness of those who share some of our genes.

To postulate any genes for altruism there is no need. As compared to the competing alleles of the same genes in organisms that, say, cannibalize their siblings or favor strangers over their own children, all the theory says is that those genes carried in organisms that contribute to their own fitness and to that of related organisms will, by definition, increase their representation in the gene pool of the next generation.

Hence, the lower the probability of shared genes-the weaker are the fitness benefits of altruism and the less effective its operation, the more distant the relationship-and. General philanthropy cuts little ice; charity, we all know, begins at home. A man can be expected to help his parents, children, his siblings, and in a pinch his cousins, nieces, uncles, aunts, and nephews.

He will extend his altruism to other members of his tribe, "race," or religion, whom he vaguely considers to be distant kin to a more limited extent. In general, what biological kin selection would lead us to expect the moral precepts of most human societies. Human altruism cannot be reduced to a blind, unconscious drive for genetic fitness.

Man is capable not only of genetic selfishness blind, but also of enlightened selfishness, a man is capable. Because we are able to distinguish the cheaters from those who play the game our way, we consciously do others good turns on the expectation of being repaid. Again, we find that cooperative behavior in man is a complex blend of genetic predispositions and cultural arrangements is complex a man's cooperative behavior.

In a new garb, inevitably, the charge will be made that sociobiology is the same old ethic of enlightened self-interest. For many people are simply incapable of dissociating a description of reality from advocacy, it will do little good to disclaim any advocacy of selfishness. It will act so as to increase its fitness.

That is what it has been selected to do, the fact remains that the social behavior of an organism is such that consciously, or more often unconsciously. In terms of inclusive fitness through kin selection or "reciprocal altruism"-for which read "enlightened self-interest over a sequence of interactions," the seeming exceptions to that rule turn out on closer analysis to be easily explainable.

Perhaps, throw some light on the study of ethics, sociobiology is not a new ethic, but it can. They will have to abandon the dogma that man is purely a product of his upbringing and his culture if social scientists want to achieve a well-rounded understanding of human behavior.

The result of both its ontogeny and its phylogeny is the lowest organism. This is also true of man. So far as we know, ethics are a cultural development and a human monopoly, but in a biological vacuum they do not exist, We are not disembodied spirits. An animal all the same but we are a very special kind of self-conscious animal. If we should forget that at some very fundamental level and we run the risk of making asses of ourselves we are mortal conglomerations of billions of cells that evolved, for the transmission of potentially immortal genes as carnal envelopes.

8.11. SOCIO-BIOLOGY: ALTRUISM AND INCLUSIVE FITNESS

To be removed temporarily from the hands of the philosophers and biologicized" Edward O. Wilson (1975) has urged that "scientists and humanists should consider together the possibility that the time has come for ethics. Sociobiology may provide the key for finding a naturalistic basis for ethics seen by Wilson, like other sociobiologists (Barash, 1977; Wilson, 1978; Alexander, 1979; Ruse, 1986).

"The systematic study of the biological basis of all forms of social behavior in all kinds of organisms" is known as sociobiology (Wilson, in the Foreword to Barash, 1977) or, "the application of evolutionary biology to social behavior" (1977, p. ix) according to Barash's concise formulation.

"To develop general laws of the evolution and biology of social behavior, which might then be extended in a disinterested manner to the study of human beings" is the main purpose (Wilson, ibidem). To discover the biological basis of human social behavior, starting from the investigation of the social behavior of animals is the ambition of the program.

In biological evolution, the sociobiologist's argument concerning normative ethics is not that the norms of morality can be grounded but rather that namely those that are consistent with the "objectives" of natural selection evolution predisposes us to accept certain moral norms.

In the social behavior of animals, it is because of this predisposition that human moral codes sanction patterns of behavior similar to those encountered. The agreement between moral codes and the goals of natural selection in social groups claimed by the sociobiologists was discovered when the theories were formulated of kin selection and reciprocal altruism.

The commandment to honor one's parents, the greater blame usually attributed to the wife's adultery than to the husband's, the incest tabu, the ban or restriction of divorce, are among the numerous ethical precepts that are also endorsed by natural selection, as has been discovered by sociobiology that are endorse behaviors.

Science and ethics belong to separate logical realms, the sociobiologists reiterate their conviction; from a determination of how things are or are not in nature that one may not infer what is morally right or wrong. They avoid committing the naturalistic fallacy in this respect.

"To devise a naturalistic description of human social behavior is to note a set of facts for further investigation, not to pass a value judgment or to deny that a great deal of the behavior can be deliberately changed if individual societies so wish" according to Wilson, (in Barash, 1977).

According to Barash (1977) puts it so: "Ethical judgments have no place in the study of human sociobiology or in any other science for that matter. What is biological is not necessarily good." What is it that evolution teaches us about normative ethics or about what we ought to do asked by Alexander (1979) and responds "Absolutely nothing"?

With the statements just quoted there is nevertheless some question as to whether the sociobiologists are always consistent. Wilson (1975), as for an example, writes that "the requirement for an evolutionary approach to ethics is self-evident. It should also be clear that no single set of moral standards can be applied to all human populations, let alone all sex-age classes within each population (Figure 8.9).

Figure 8.9. Altruism–bees.

Source: Image by Wikimedia commons.

To impose a uniform code is therefore to create complex, intractable moral dilemmas." Moral pluralism is, for Wilson, "innate." Biology, to decide that certain moral codes then, with human nature and therefore unacceptable, helps us at the very least (e.g., all those pretending to be universally applicable) are incompatible.

In favor of the biological determinism of ethical norms, it does approach determinism from the negative side but this is not quite an argument: that some are not compatible with biological nature because the range of valid moral codes is delimited by the claim.

However, Wilson goes, further when he writes: "Human behavior–like the deepest capacities for emotional response which drive and guide it–is the circuitous technique by which human genetic material has been and will be kept intact. Morality has no other demonstratable ultimate function" (Wilson, 1978).

How is one to interpret this statement? why ethical behavior exists at all, it is possible that Wilson is simply giving the reason; to evaluate morally their actions as a means to preserve their genes, their biological nature his proposition would be that humans are prompted. But this proposition is erroneous.

Because of their innate ability for anticipating the consequences of their actions human beings are by nature ethical beings in the sense I have expounded earlier: they judge morally their actions, for formulating value judgments, and for free choice.

To preserve their genes or serve any other purpose, human beings exhibit ethical behavior by nature and necessity, rather than because such behavior would help.

8.12. IMPLICATIONS OF SOCIO-BIOLOGY FOR MORAL EDUCATION

In moral psychology, studies in sociobiology and its related fields have provided important lessons on emotional factors. In the evolution of humankind, emotions have a long history and when working with moral dilemmas they do not require consciousness, in moral thinking which makes them powerful elements.

We should concentrate upon lessons from sociobiological studies and apply their implications to school classes to improve moral education in schools. To theories of moral psychology Wilson (1999) tried to apply studies of sociobiology. As for an example, with his own theory he tried to explain the Kohlbergian model.

As a biological adaptation can help us decipher the meaning of moral canons, he insisted that interpreting the activity of emotive centers. He argued that moral development is only a more complicated and less tractable version of the genetic variance problem, by affirming this.

To explain the various Kohlbergian stages Wilson presented several examples—for instance, to perform altruistic acts based on personal principle (Stage 2) the selective advantage for young children to be self-centered and relatively disinclined, and the adolescent's tendency to be more tightly bound by age-peer bonds within their own sex and hence hypersensitive to peer approval (Stage 3) (Kohlberg, 1981, 1984; Wilson, 1982).

By natural science Wilson (1999) also argued strongly that normative-moral decision making can be fully analyzed and predicted.

8.13. THE ETHICAL IMPLICATIONS OF EVOLUTIONARY THEORY

Because the arguments had become deadlocked and other, more pressing issues emerged for decades, the issue of evolution's impact on ethical theories had been understandably shelved. It has been brought back out and examined with renewed vigor, however, now with major advances in biology, especially sociobiology and evolutionary biology (Figure 8.10).

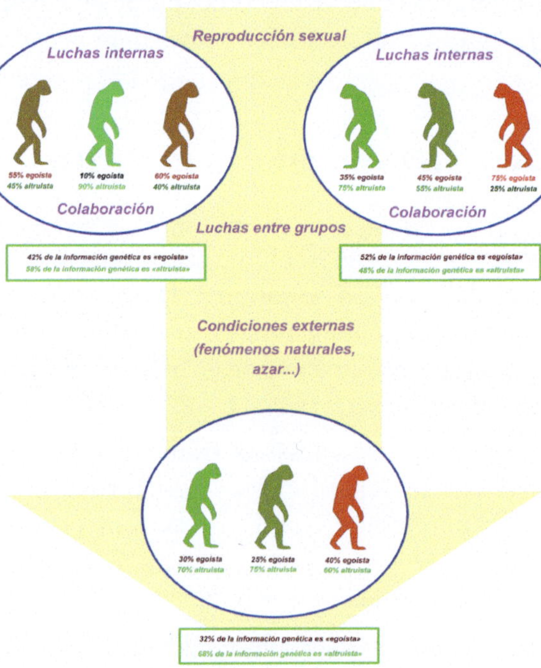

Figure 8.10. Dawkins' evolution theory.

Source: Image by Wikimedia commons.

Evolution and ethics surely deserve a second go around also, because Darwin's theory makes up such a large part of the way most of us view our species and the world. To introduce the basics of evolutionary moral theory, we read from William F. Quillen, Jr. Ethics can be placed for the first time upon a scientific basis, the contention of Evolutionary Naturalism is that by the application of the theory of evolution to the investigation of moral phenomena.

By that same process of natural selection which has determined the development of the present structure of animals, including man, from some earlier form it is supposed that moral sentiments, customs, and judgments have been gradually developed over a long period of time. The central theme of evolutionary ethics from Darwin through the present generally described by this statement.

A more recent evolutionary naturalist Richard D. Alexander, has written that "it is necessary to understand the life interests–therefore the life patterns–of humans as outcomes of an evolution guided principally by natural selection," and that by "understanding the evolutionary significance of the human organism, and the nature of individuality of its evolved interests, we may derive useful insights into human concerns about morality."

To identify some of these insights what follows now is an attempt, and will put forth three specific ones–the body as a compass, the need for adaptability and the importance of natural motives. What is involved in evolutionary ethics before that, however, there will be an exposition, the traditional debate around it, and my perceived limitations of the position.

To reduce human nature to biology alone, first it is important to make clear that evolutionary ethicists do not attempt. Matthew H. Nicki does not cite in an essay on the subject, but biology is not easily reducible to chemistry and physics as he mentions C. G. Hempel's assertion from Philosophy of Natural Science.

He says, "[f]or example, you may resolve penicillin to a chemical equation, but not as a substance produced by a living fungus Penicillium notatum." To reduce morality to biology. Morality never involves the same circumstances Nitecki argues that it is even more difficult, which biology cannot describe at the moment so its components are things such as intentions, beliefs, and wants.

To reduce morality simply to genes is not attempted by the evolutionary ethics. To be moral does not caused by the genes. We are not, as for an instance, genetically coded to refrain from lying. Environment is also as much of a force upon our behavior as our genetic makeup. "Evolution and Ethics," is largely a response to this lecture by Huxley according to John Dewey's essay.

Human morality does not contradict nature, or even another part of nature, but that it expands the possibilities of nature is the underlying idea by Dewey's. Nothing ever battles nature for Dewey, but, instead, by means of human artistic endeavor nature sculpts and fiddles with itself, even sometimes. Nature is all-encompassing. With Huxley's interpretation of the garden and the gardener he tases issue.

In a certain way that results in what we call a garden, to Dewey, the gardener is actually a component of nature, a natural entity, modifying other components. The seeds came from another place in nature but the gardener may have planted seeds that would not normally have been in that spot (Figure 8.11).

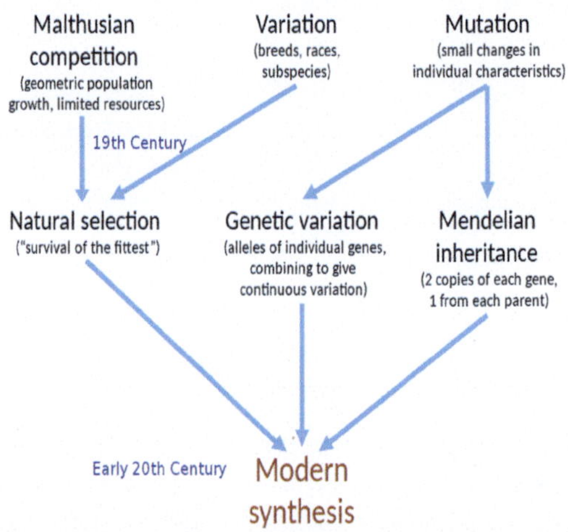

Figure 8.11. Diagram of the ideas brought together in the 'modern synthesis' in evolutionary biology of the early 20th century.

Source: Image by Wikimedia commons.

He says that the gardener will modify the amounts of sun and water that reach the area, but that these things still "fall within the wont and use of nature as a whole." The gardener must keep up with the forces trying to break down his or her work admitted by Dewey. It is a struggle. With his current conditions, the gardener's struggle, though, is not with the whole cosmos. This is where the analogy is misapplied by Huxley. To grow plants truly is an adaptation to conditions is the ability. Morality is an adaptation to conditions to maintain the metaphor. It would not know how to argue that bees are not also battling nature, if growing a garden conflict with nature.

8.14. CONCLUSION

In the conclusion of the chapter, it discussed about the ethical implications of sociobiology. It also discussed about the sociobiology and the origins of ethics. In this chapter, several moral evaluations and moral norms have been discussed. This chapter also discussed about the necessary and sufficient conditions for ethical behavior.

Towards the end of the chapter, it discussed about the evolution of information processing and question of ethical ethics. It also provides highlights on the moral norms for religious and evolutionary proposals. In this chapter, implications of sociobiology for moral education, and the ethical implication of evolutionary theory have also been discussed.

REFERENCES

1. Ayala, F., (1987). The biological roots of morality. *Biology & Philosophy, 2*(3), 235–252. [online] Available at: https://escholarship.org/content/qt85j4s12p/qt85j4s12p.pdf?t=pd1n2n (accessed on 27 June 2022).
2. Jeon, C., & Han, H., (n.d). *Some potential Contributions of Sociobiology to Moral Psychology*. [e-Book] . Available at: https://s-space.snu.ac.kr/bitstream/10371/72998/1/vol19_7.pdf (accessed on 27 June 2022).
3. MUÑOZ-RUBIO, J., (2004). *Sociobiology and the Origins of Ethics*. [online] Ludus-vitalis.org. Available at: http://ludus-vitalis.org/html/textos/22/22-05_munnoz.pdf (accessed on 27 June 2022).
4. Quillian, W., (2009). *The Ethical Implications of Evolutionary Theory* (2nd edn.). [e-Book] https://openjournals.bsu.edu. Available at: https://openjournals.bsu.edu/stance/article/download/1729/1047/ (accessed on 27 June 2022).
5. Van, D. B. P., (n.d). *Sociobiology, Dogma, and Ethics*. [e-Book] http://archive.wilsonquarterly.com/. Available at: http://archive.wilsonquarterly.com/sites/default/files/articles/WQ_VOL1_SU_1977_Article_03_1_2.pdf (accessed on 27 June 2022).

INDEX

F

G